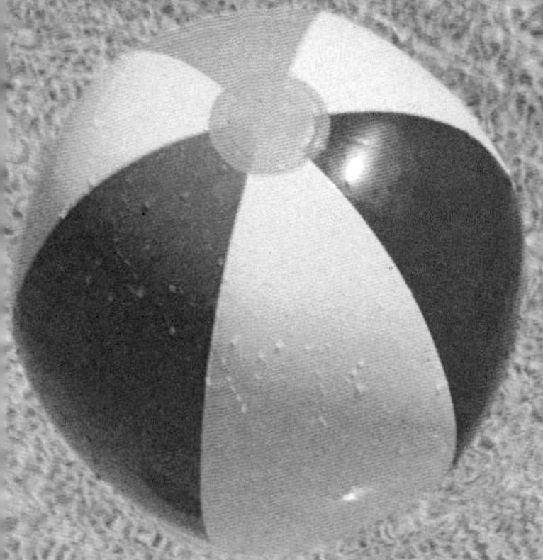

W9-BFS-725

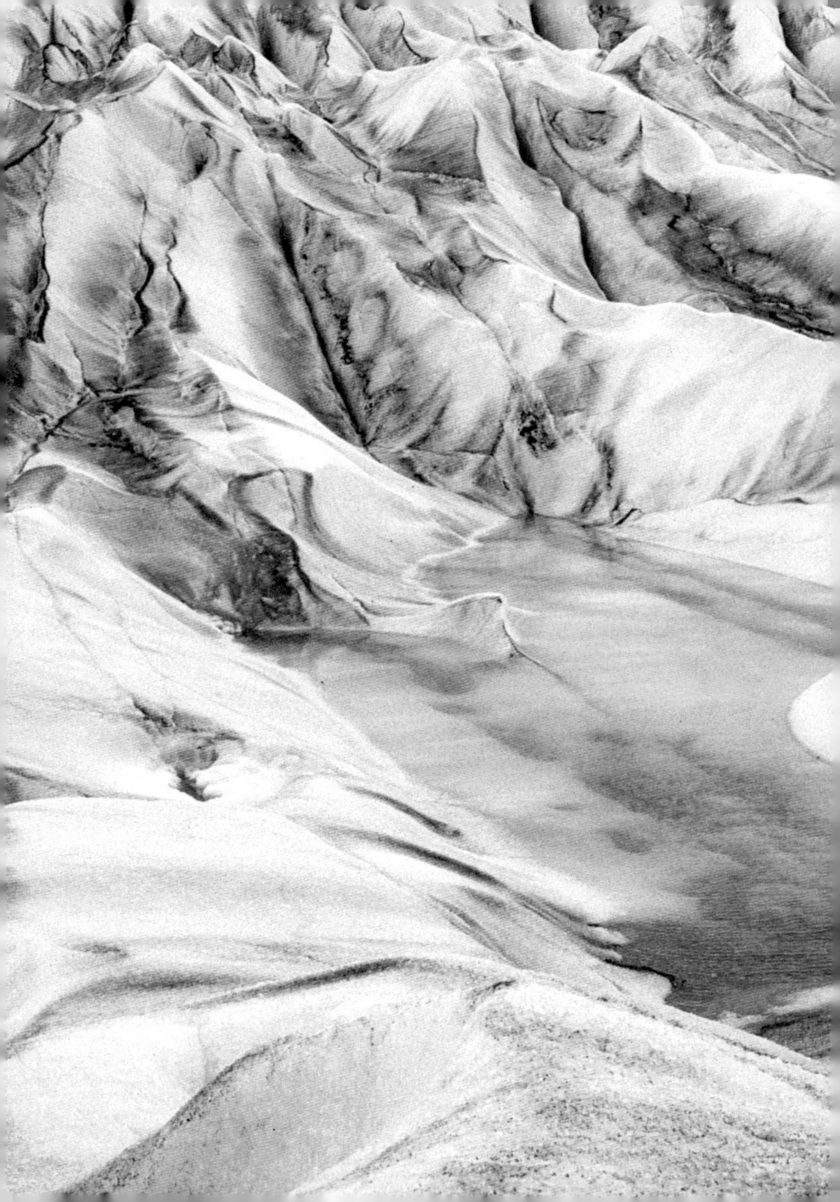

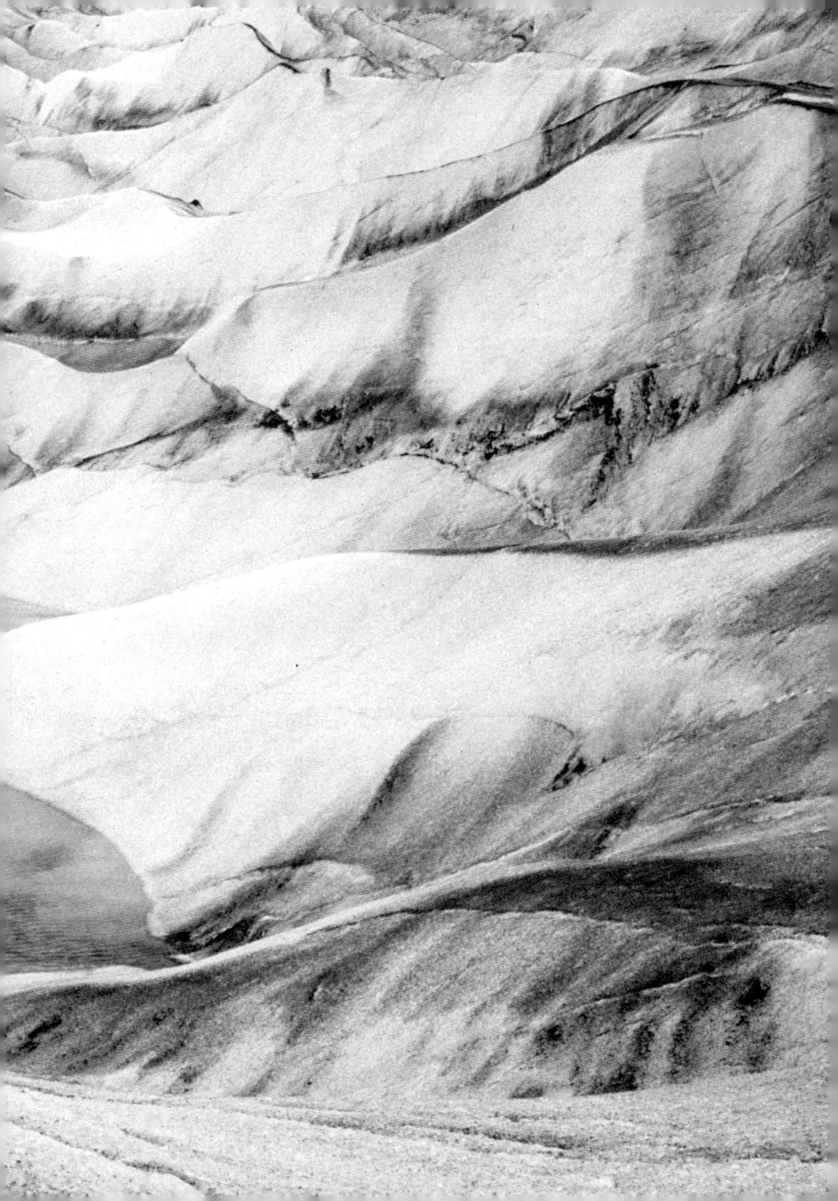

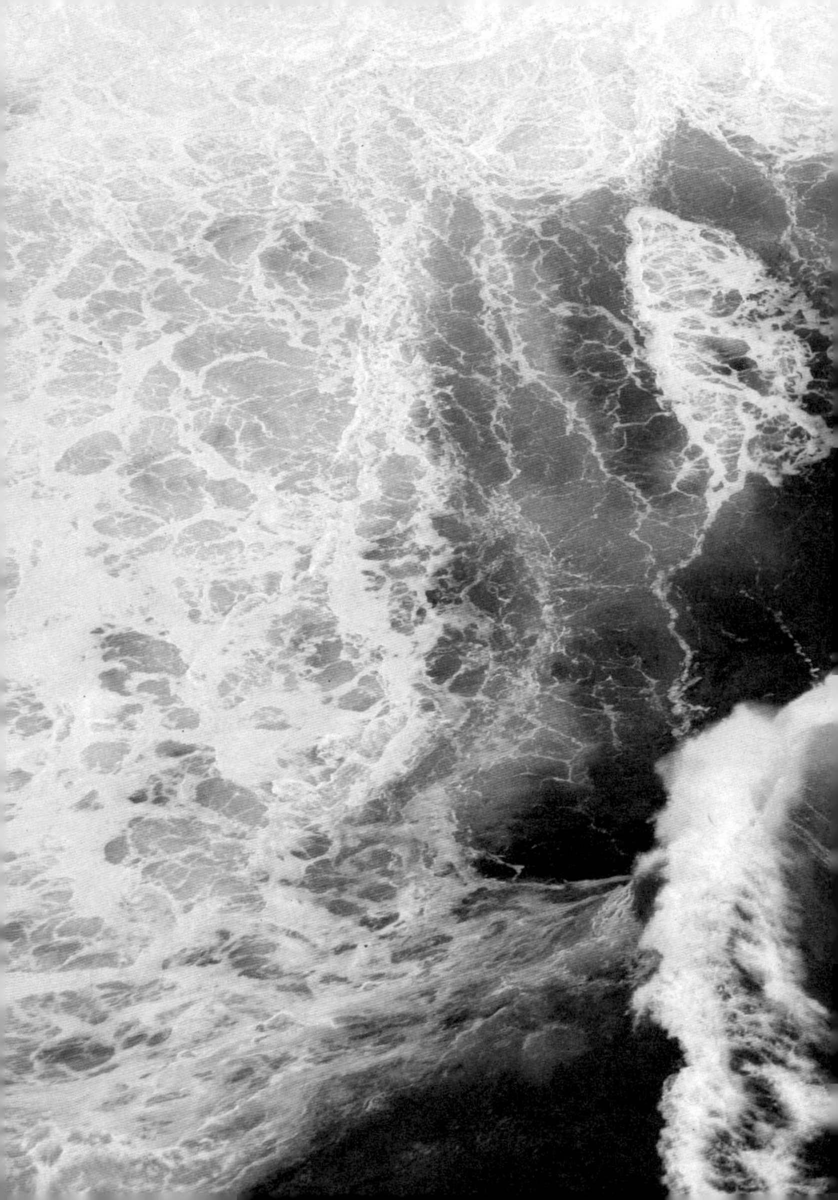

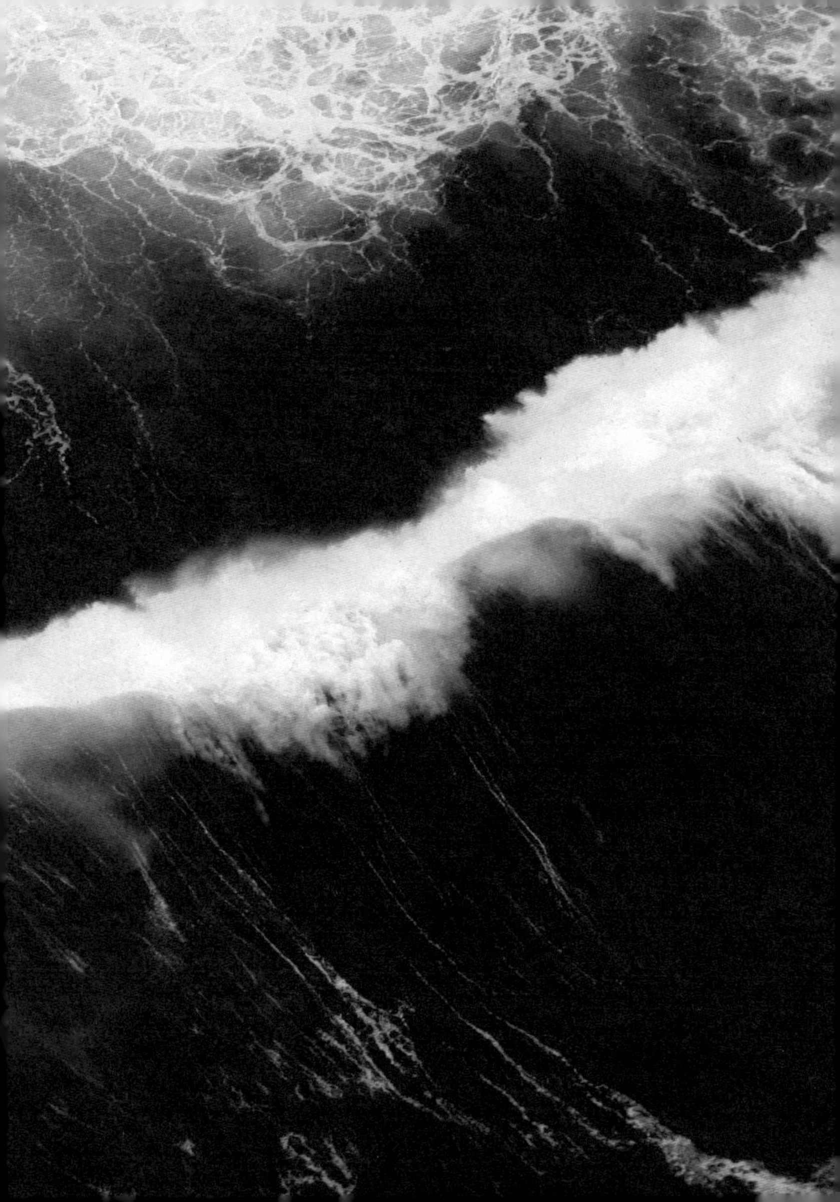

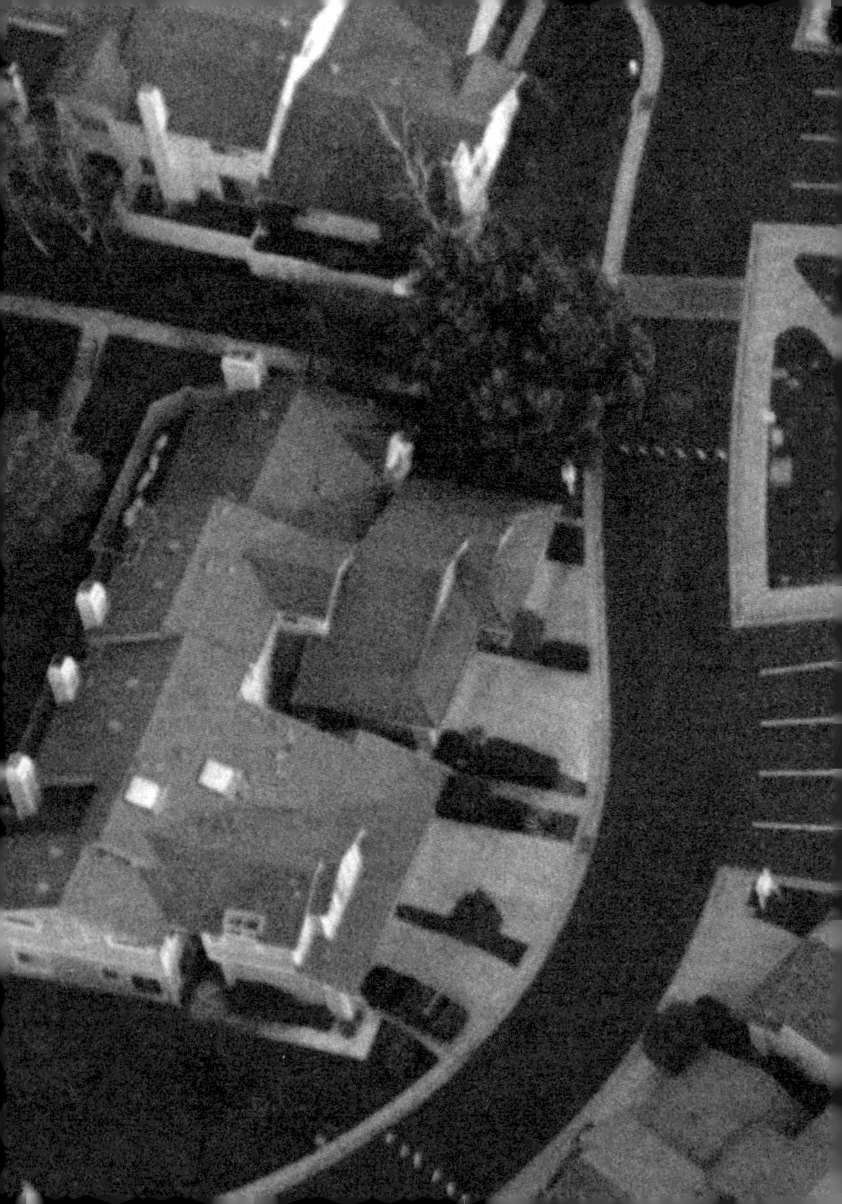

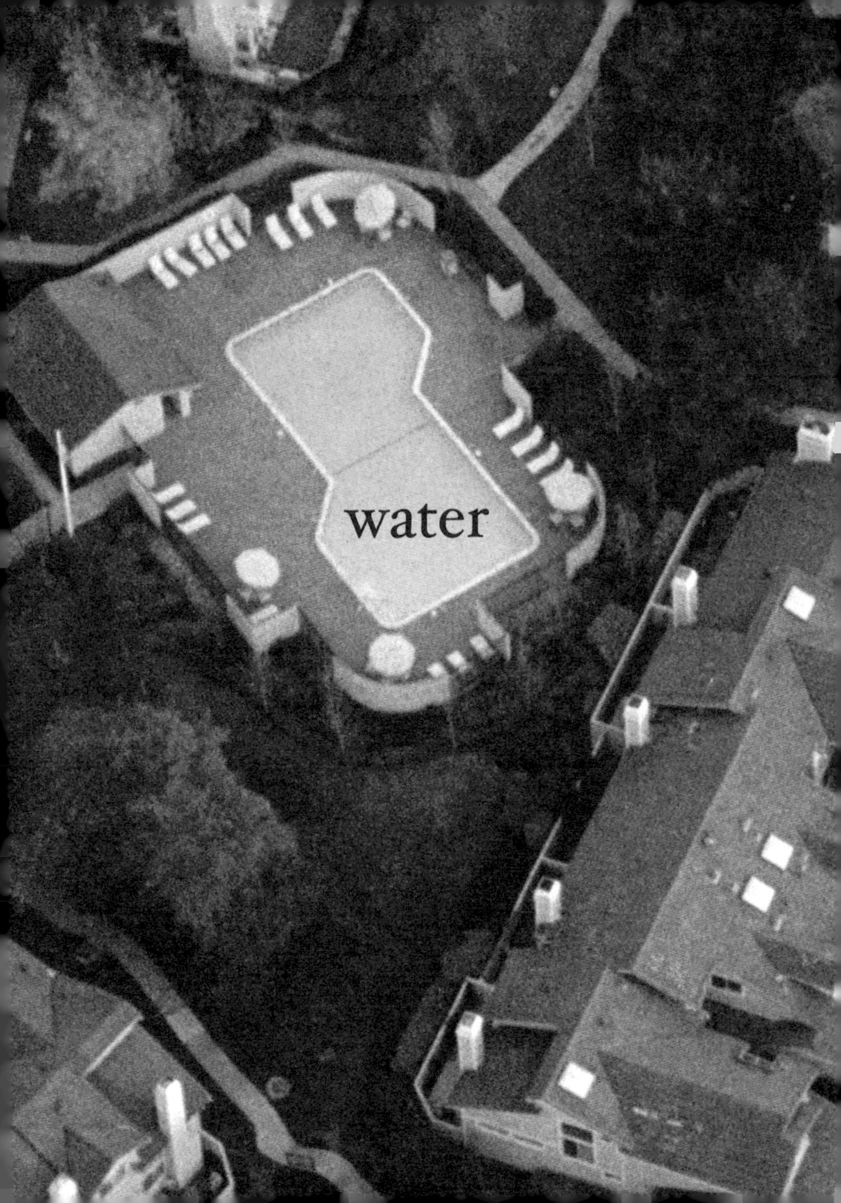
water

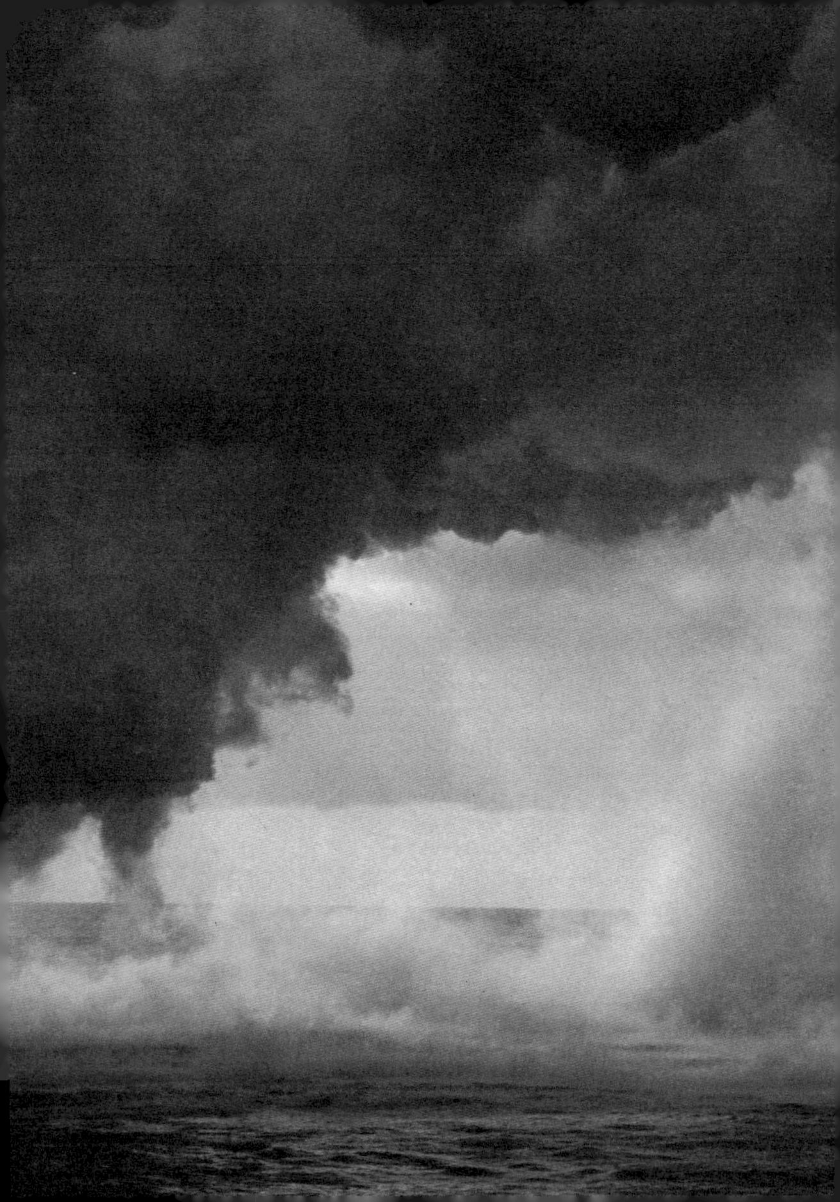

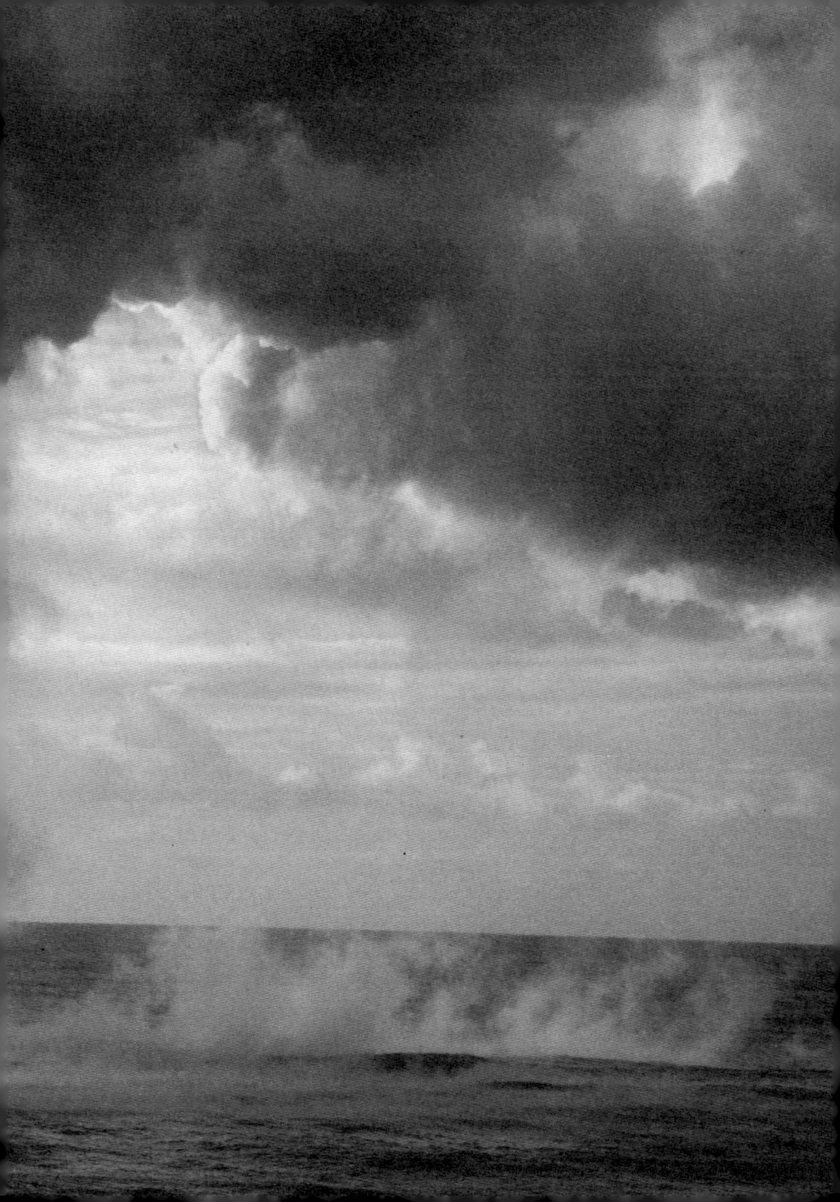

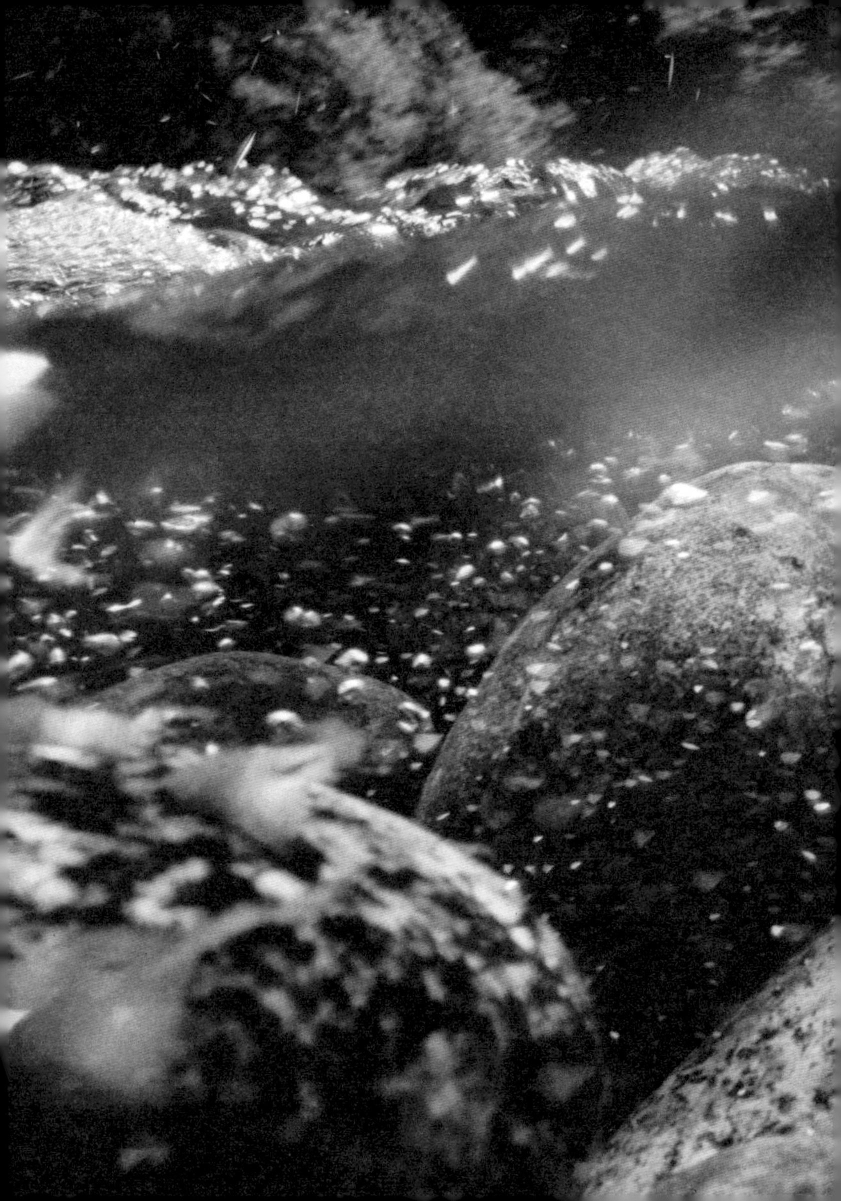

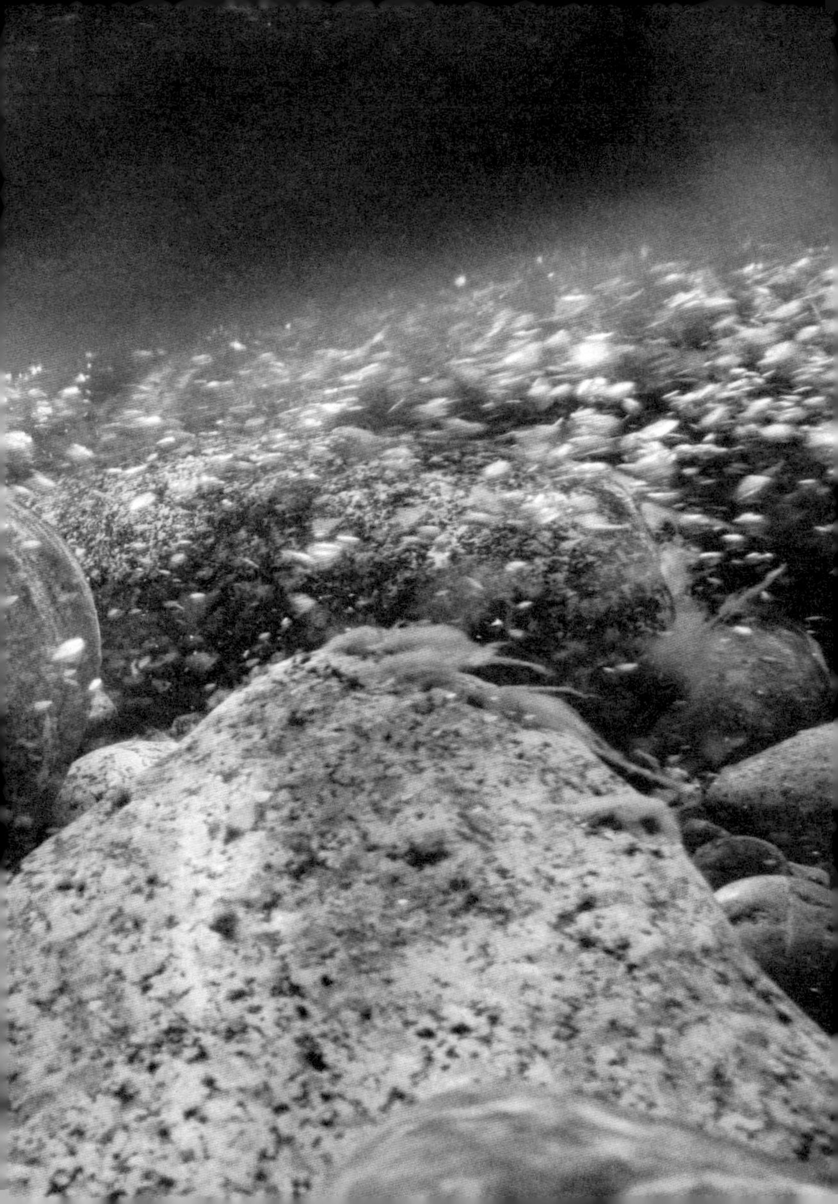

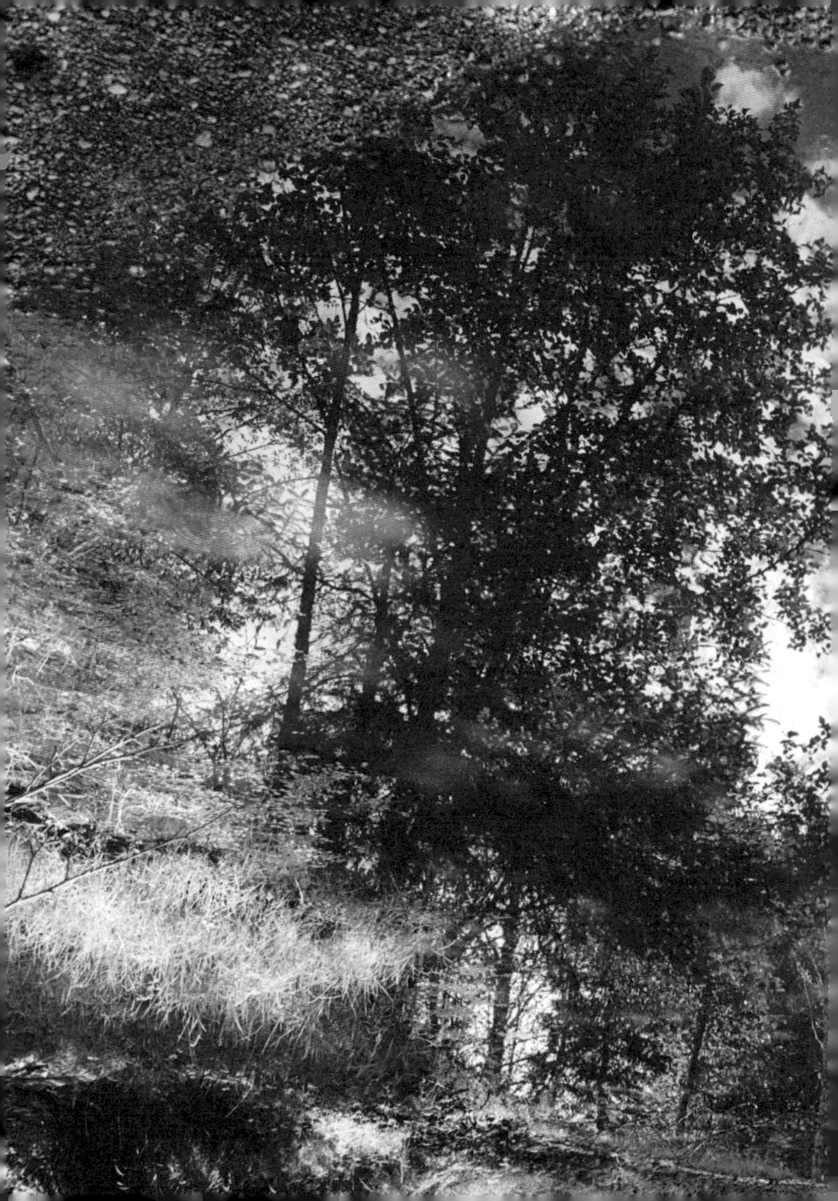

water

ALPHABET CITY NO. 14 Edited by John Knechtel

An **Alphabet City Media** Book
The MIT Press Cambridge, Massachusetts, and London, England

MIT Press books may be purchased at special quantity discounts for business or sales promotional use. For information, please email special_sales@mitpress.mit.edu or write to Special Sales Department, The MIT Press, 55 Hayward Street, Cambridge MA 02142.

This book was set in Benton Sans and Hoefler Text.
Designed by Studio Blackwell.
Printed and bound in Canada.

ISBN: 978-0-262-01329-1
ISSN: 1183-8086

10 9 8 7 6 5 4 3 2 1

Table of contents

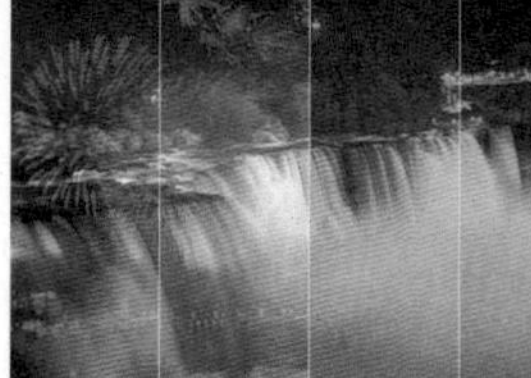
40

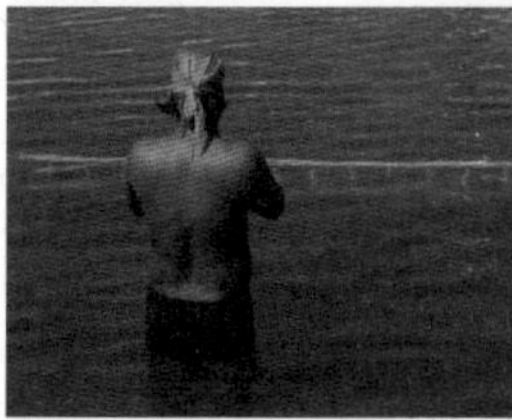
60

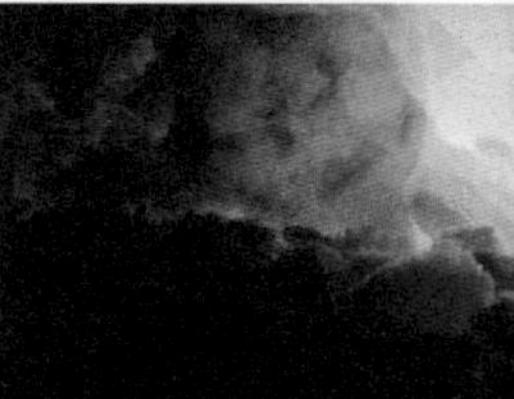
72

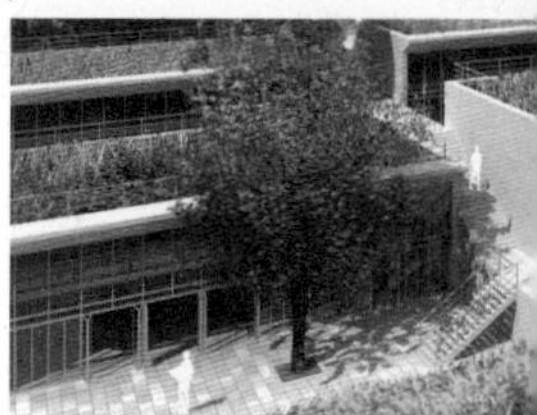
92

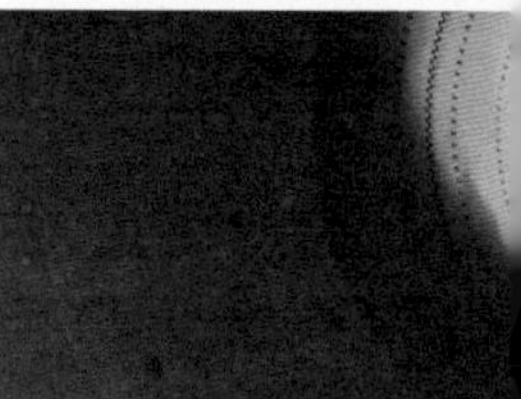
102

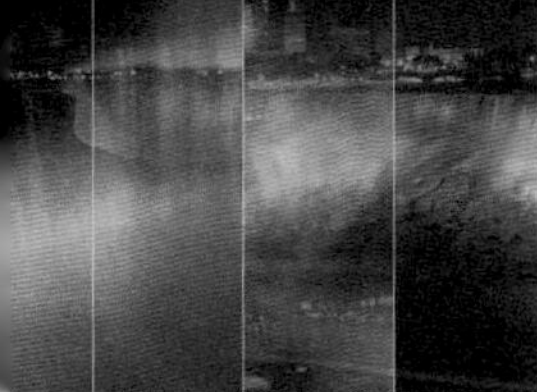

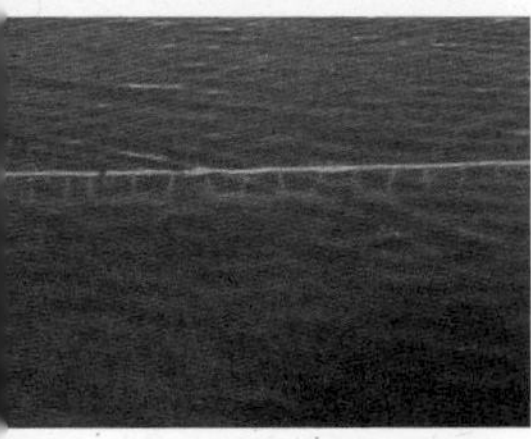

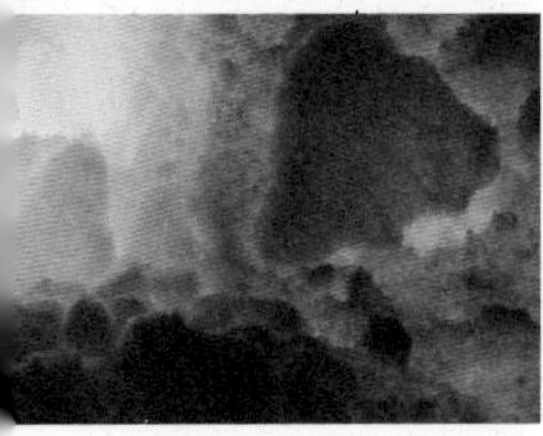

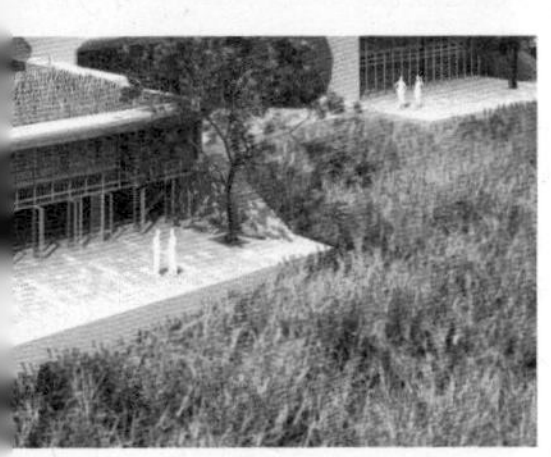

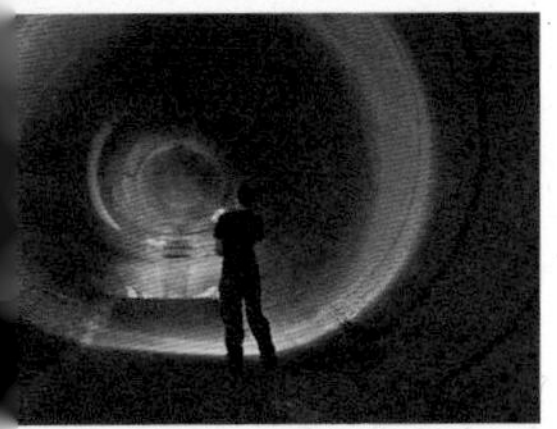

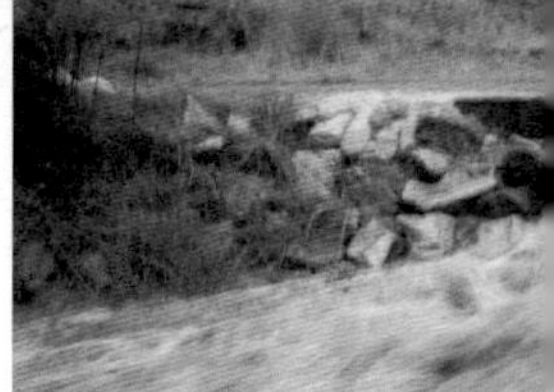

114

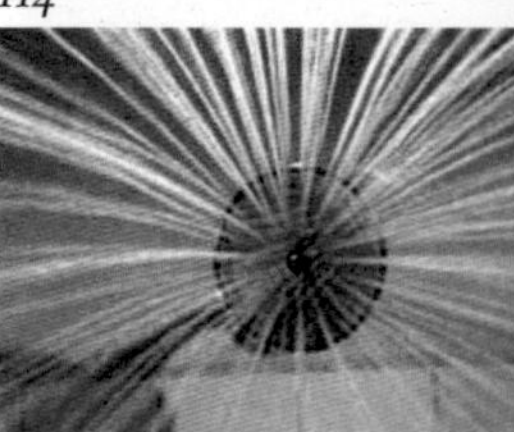

162

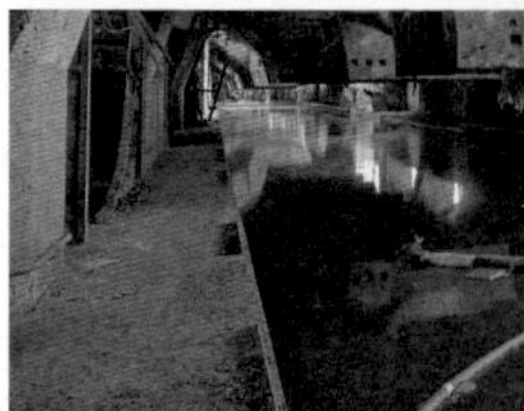

206

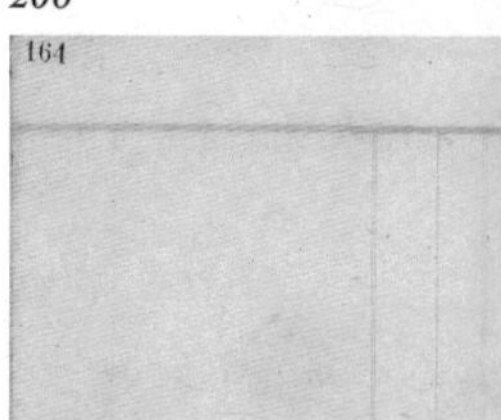

206

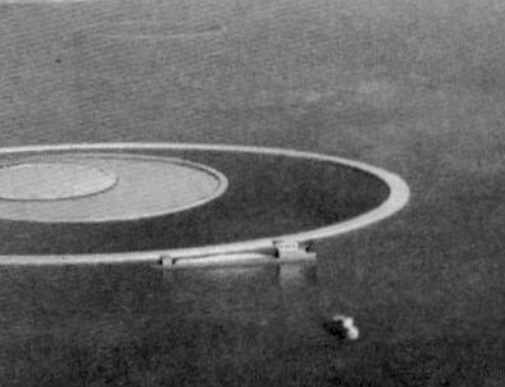

280

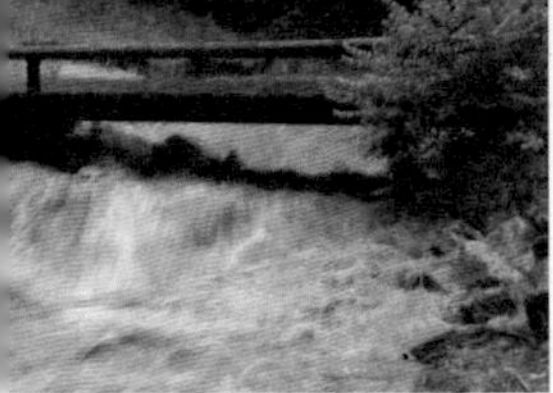

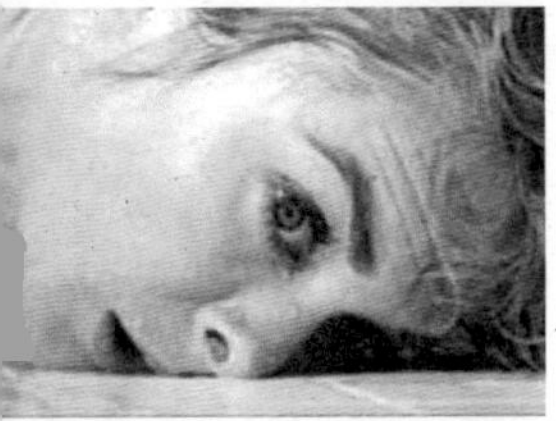

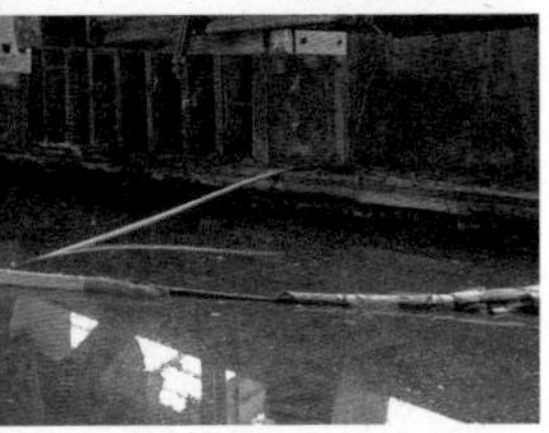

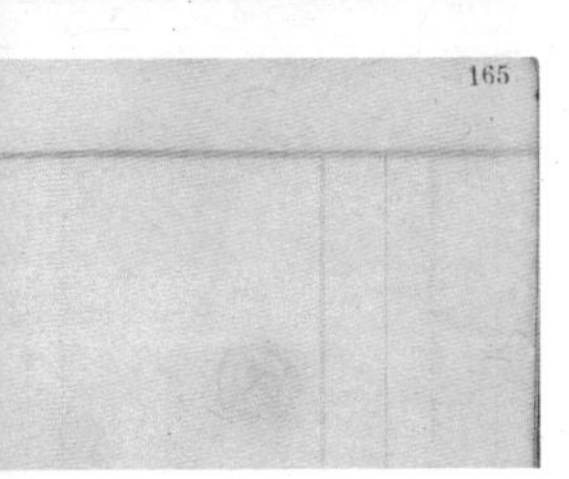
165

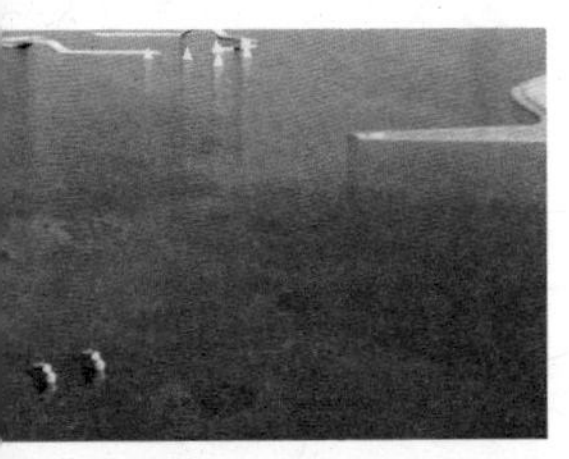

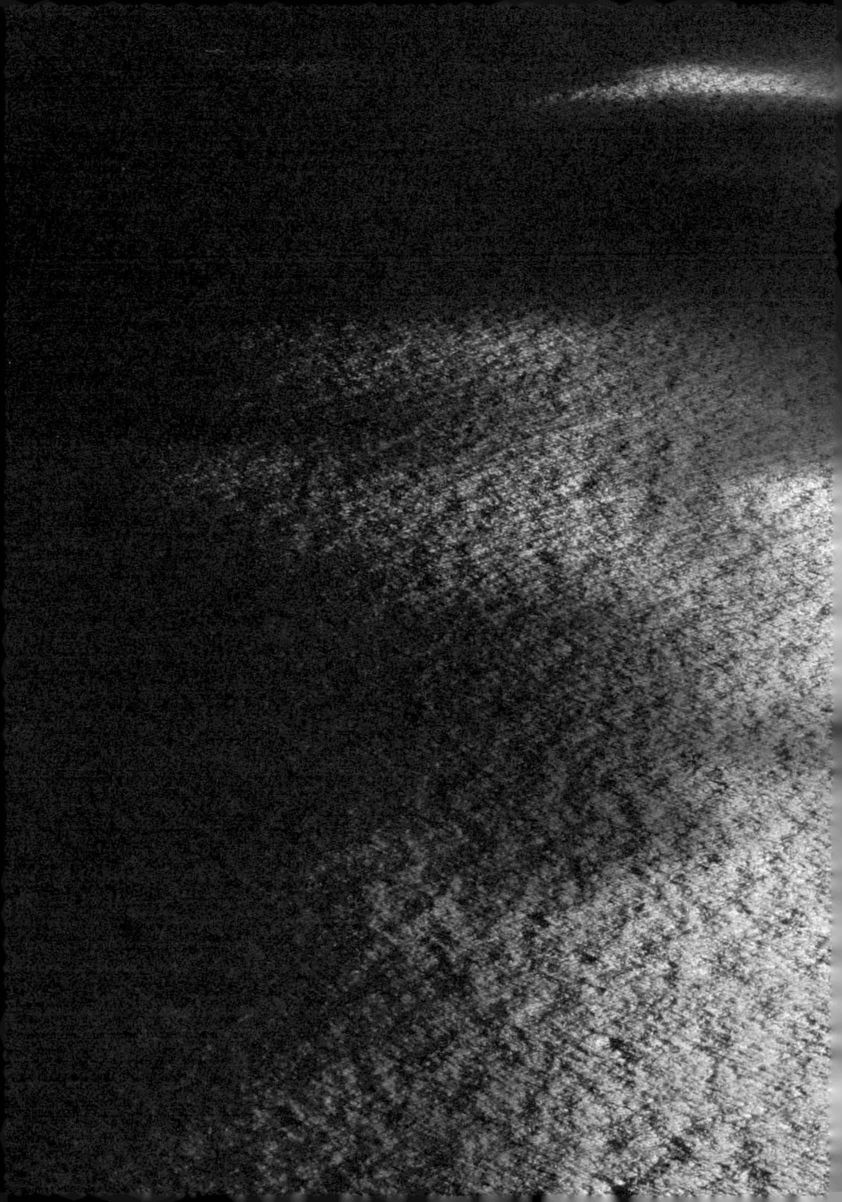

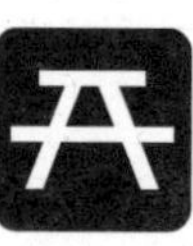

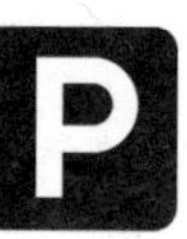
P
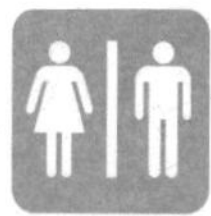

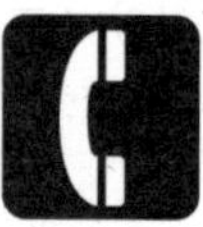

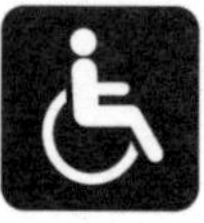

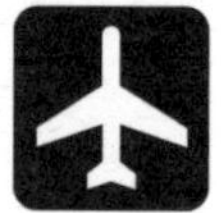

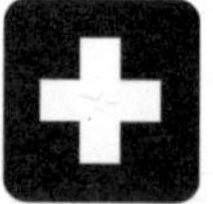

Water City After years of living in the core of this metropolis I moved to the lakeshore, and only then did I begin to properly understand Toronto as a water city. From the balcony of my sixth-floor apartment I can observe thousands of cubic kilometers of shore and water and sky. Standing here, I become an accidental hydrologist in my own small research station. On sunny days, when the water is brushed by wind and cloud, the lake peacocks bright cobalts and ceruleans, streaked with slate and deep azure. There are days when masses of cumulo-nimbus conquer the city from the southwest, drenching my world as they pass. Then there are the silent days when the clouds enclose my balcony in a soft wadding of mizzled French gray. And on very rare days, when the first bitter winter winds hit the warm lake water, hundreds of vapor plumes shoot up high above the surface.

For an hour or two, the border country between lake and cloud is feathered and unstable. Then the sky clears and the horizon returns once again—the hard wet edge of the planet's curving rim.

The scale of what I can see gives me the measure of what I cannot: the invisible depths, the topography of the lakebed, and the wet shadow of groundwater below that. Invisible as well are the ancient aquifers running beneath the city. They are fed by rain, by snowmelt, by underground reserves—all gradually filtered through the layers of clay, sand, gravel, and rock that make up the ground here.

But this great cycle of water's transformations is interrupted by the city's hard skin, an impermeable membrane of roofs and roads, parking lots and laneways, tennis courts and playgrounds. Rain streams off this city-skin: channeled, controlled, eventually swept along storm sewers and into my lake, bringing a plague of evils as it rushes along. "A lake is a sump," says Julian Caldecott in his survey *Water*; everything in its drainage basin flows in one direction, much of it sinking to the lakebed in a permanent, toxic sediment. Caldecott describes some of the contents of this urban wash: "agricultural chemicals, sewage, the oily effluent of car washes, the soapy traces of household detergent, and the accumulated trash of street litter."

All cities foul their water. In *Invisible Cities*, Italo Calvino demonstrates this fate through his imaginary city of Armilla which is nothing but its plumbing: "It has no walls, no ceilings, no floors: it has nothing that makes it seem a city, except the water pipes that rise vertically where the houses should be and spread out horizontally where the floors should be: a forest of pipes that end in taps, showers, spouts, overflows." Our urban water systems were built to prevent cholera and other water-borne scourges, but their unintended legacy has been the long-term degradation of our cities' water ecology. The planet now has Armillas clamped to every corner of its surface. It will take a century and more to rescind their dirty works.

This story could have been different. The then-young Los Angeles, for example, could have followed Frederick Law Olmsted and Harlan

Bartholemew's nineteenth-century plan for a green and hydrodynamically balanced city, setting aside large tracts of land for public amenities including "greatly elongated parks several thousand feet in width" running along natural floodways. Their plan would have delivered a radically different LA from the water-starved, privatized city it is today. But the public interest was no match for the interests supporting unfettered private property rights, and the latter won, decisively. LA's waterways were buried under expensive and disaster-prone real estate.

LA will not be unmade; the development of urban form cannot go in reverse. But we can leverage the progress water policy has made in recent decades and a multitude of promising technologies to design new water cities, cities whose water commons, as they leave the man-made systems, are cleaner than they were when they entered. Achieving this result will require an assemblage of approaches, from local water-storage projects in developing countries to protective greenbelts of undeveloped land wrapping urban centers. We'll have to plant toxin-absorbing vegetation to break down salt and oil, and pave our roads with surfaces that filter water instead of repelling it. The standard rooftop will be a garden, and every house will have a rainwater recycling system. Above all, to build water cities that improve rather than degrade their water supply, we will need a political culture with the imagination to invest in the long-term public interest.

From my balcony, I see only the small eddy Toronto makes in the global water cycle that surrounds us. This powerful and almost invisibly slow process can easily be ignored, as can its steady corruption by our cities. But ignoring it is precisely what will cause my city, and others, to slip, fold, and buckle at the whim of water. In this volume we explore what could happen if we embraced our watery nature, and learned to care for this essential life-sustaining liquid.

—JOHN KNECHTEL

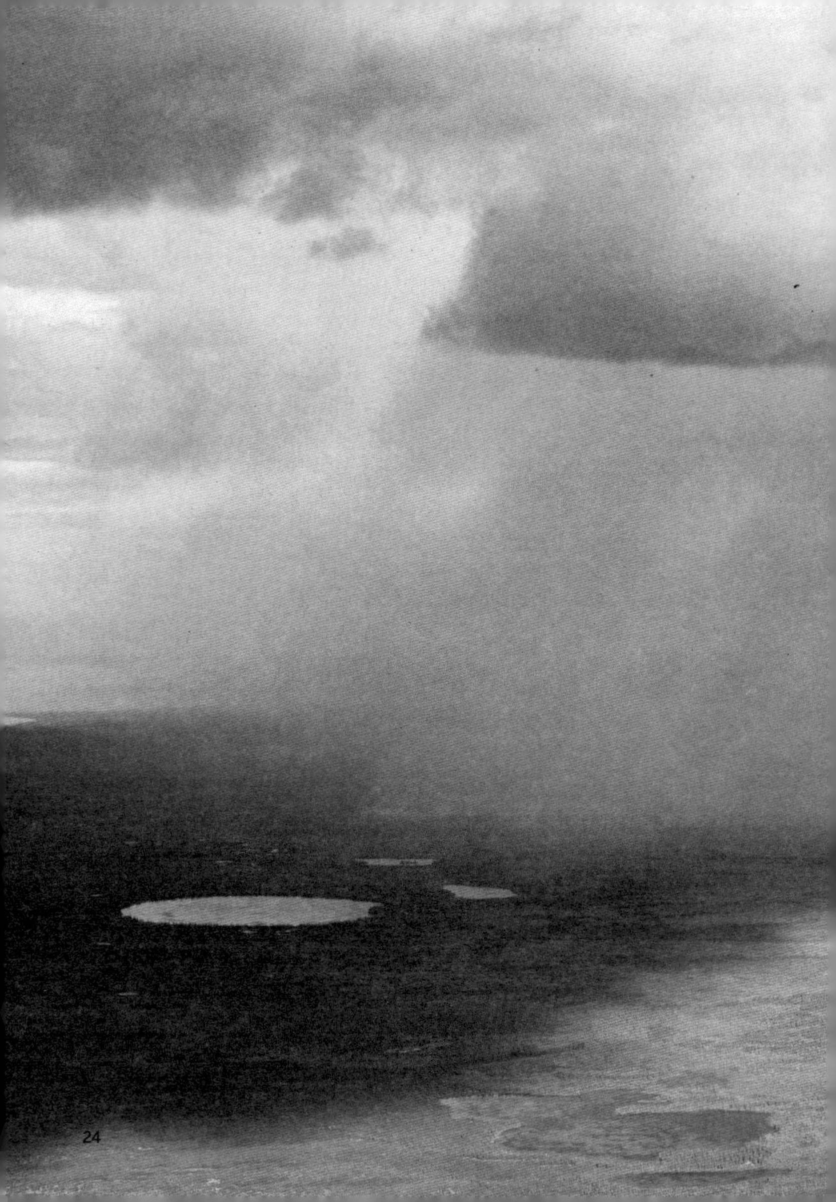

The Waters of Metaphysics *Timothy Stock*

"Water is everything."—So said the first Greek philosopher, Thales. Water is storied in spiritual traditions throughout the world, and has ritual significance in every religion. How right, then, that the core questions of metaphysics—cosmology, identity, modality, and being (ontology) can all be understood through careful attentiveness to water. The oceans of the earth feed the clouds, the clouds precipitate and freeze, and the meltwater feeds the rivers of the earth, which in turn run to the sea. Water behaves like being: endlessly changing, yet ever the same.

> *All the philosophers ... may be marshaled in one line ... "Oceanus the origin of the gods" ... [as] all things are the offspring of flow and motion.*
>
> —PLATO *THEAETETUS 152E*

In 1992 thousands of rubber ducks fell off a boat in the middle of the Pacific and washed up on the Alaskan shore. Three years later they started to appear in Japan, Hawaii, and Australia. After eleven years they appeared in Massachusetts following a long, slow migration under the Arctic ice. And in fifteen years they had reached the UK, on the opposite side of the world. The Greeks thought of the Ocean as the largest river, one that circumscribed the world. The ocean was the birthplace of the gods, and the deep, subterranean element to, from, and through which all things were thought to move. In a sense, they were right. The ocean is water in flux and change to a sublime degree, but is also the continuity of currents and the origin of the trade winds. It feeds the earth with its gifts while at the same time ravaging it with storms. Even Poseidon, the mighty and changeable god of the Mediterranean, was a weak shadow of the great Oceanus.

Philosophers today commonly misconstrue the river as the ideal of change and flux for the Greeks, largely because of Heraclitus's famous (and infamously misquoted) dictum "You cannot stand in the same river twice." But rivers are hardly changeable in comparison to the vast ocean. A river may change its course, flood its banks, or dry up in drought, but each of these events is a reversal of our expectations, a catastrophe. The ocean is flux on a cosmological scale; it is that into which and out of which all things go. Modern ecology tells us that all elements eventually move throughout the ocean and thereby through the food chain. And according to evolutionary theory it was these elements—the salt, detritus, and waste dissolved in the vast ocean—that struck the spark of life for all species. Just as Plato saw fertile "flow and motion" in the Titanic god Oceanus, we now see the ocean as the solution upon which all life depends, and the dissolution into which all life returns; all being, becoming, all becoming, being.

Cosmology is the study of the world in its totality, and the Ocean is a venerable cosmological metaphor. It is the "purest and impurest water," according to Heraclitus. In a simple sense, this is a reflection on the fact that for the creatures of the sea the chemical composition of the ocean is ideal, whereas for others, like us, it is hazardous and even fatal to drink. More subtly, the oceanic paradox teaches us that purity is relative, at least at the species level, and that "pure water" would be an ecological curse. Yet this is perhaps the most important way that the ocean teaches us about the substance of being. To speak of one unified substance is always to speak ambiguously about purity and impurity. We can speak of cosmology as being in its unity, but such unity is only comprehensible as a unity of diverse forms. In this sense water behaves like being: salt, brackish, and fresh, it surrounds us, saturates our world, and provides life or death to us all, depending on its solution.

RIVER/IDENTITY

> *Upon those who step into the same rivers, different and again different waters flow.*
>
> —HERACLITUS, *FRAGMENT 12*

Yet even the infinite ocean must be distinguished from the water that makes it up, just as in metaphysics one distinguishes between speaking about being as a whole, and speaking about the forms of being. Knowing that all things flow from the ocean, the whole, the totality, is markedly different from knowing the twisting paths it traverses. These twisting paths, local ecologies, and different peoples are best understood through the waters of the earth's rivers. The children of the Titan Oceanus were the innumerable gods of rivers, and their significance is visible in their name: the Potamoi, related to our word "potable." The rivers are those waters that affect humanity most proximally, they are the waters that quench our thirst.

We identify ourselves by rivers. The borders of states and nations, of cultures ancient and modern, are all tied to rivers. The Fertile Crescent, the Nile, the Yangtze, and the Mississippi all serve as the agricultural and economic impetus for a people to flourish, to come into being. And peoples are defined and redefined by rivers in their dominion; consider the disputes of the various factions that fought back and forth across the Rhine: German and French, but also Roman, Celt. The banks of a river stand as the most stable of national boundaries, as well as a reminder of the catastrophic floods of humanity that can overflow borders in times of crisis and ambition.

And yet a river does not need war to shape the identity of its people. The Yellow River, also named "China's Sorrow," speaks to the grand scale of a river's impact on human life. Its fertile soil deposits have fed the Han people for more than four millennia, and yet the depth and breadth, indeed the very damp richness of these deposits, have also caused it sometimes to flood or catastrophically change its course. This steady stream has nurtured the Han throughout their long history, save for the few unlucky generations that were swept away when diversions in the river wreaked near-total destruction.

Like all political and economic arrangements, life on the river is both fruitful and fragile. The Zen Master Shenshan told his dharma brother Dongshan, "If I make a mistake in my steps, then I won't live to cross the river." In reflecting on the episode the Zen Roshi John Loori observes, "the river treats all things equally." This captures nicely the paradox often noted by metaphysicians, that the river represents both the stability and the unpredictability of being. As I mentioned earlier, Heraclitus is often misunderstood in having said "one cannot step in the same river twice," most notably by Cratylus who one-upped his teacher to state that one could not even do it once. But Heraclitus merely stated that the river is the metaphysical problem of identity:

while the river is the same, the water is ever changing. We have learned that we must take great care to protect our fragile river ecologies, but the river has more to teach us about the precarious balance between prosperity and destruction. The river shows us that our world is not simply stable and not simply fluid, but that when we take a step into the world, we must step into one or the other, stability or fluidity: for our own benefit, at our own peril.

CLOUDS/MODALITY

> *Strepsiades: Come on, Clouds, give me some good advice. ...*
> *Clouds: Lap it up—make haste/get everything that you can raise./Such chances tend to change and turn/into a different case.*
>
> —ARISTOPHANES, *THE CLOUDS*

There is no greater metaphor for the caprice of our world than the caprice of the cloud—weightless, ephemeral, ever-changing. Aristophanes used the image to ridicule the philosophers in their shifting vagaries. Clouds are the bearers of fortune, opportunity, and chance. Physicists will sometimes speak of "clouds of modalities," for example, to express the phenomenon of electrons moving about the nucleus of an atom. One cannot speak of the electron's position and velocity accurately, but must speak of the different modes, or valences, of its activities. The cloud is vagueness, the coexistence of possibility and necessity at one and the same moment.

Who hasn't sat idle on a summer afternoon, simply to search for shapes in the clouds? The cloud can become anything, can adapt to any form of being, because of the simplicity of its material structure. And yet each cloud coheres with itself; it always looks like something—"giants, animals, ships and castles," as Tomie dePaola described them in his Cloud Book. The waters of a cloud show us the indifference of being to form, as well as our constant need to give form to being. And yet any form we give the cloud is simply one possibility; we blink, and the vapors have shifted, become new, become something other. Modality, the study of possibilities and necessities, has always been informed by the ancient observation that only by exploring fortune, opportunity, and chance can we understand the endless flux of time that carries us forward. The clouds are one elegant demonstration of the conflux of mode and time. As one shifts, so too, almost imperceptibly, does the other.

Hence, the careful observation of clouds is not simply idleness. Clouds tell us as much of the future as we can know: when and how the rains will come, whether there will be warmth or cold, storm or calm. In our childhood games with the clouds we learn water's central lesson, that the heart of all prudence lies in attentiveness to the changing forms of our world, and it is upon this attentiveness that our flourishing rests.

HAILSTONE/ONTOLOGY

> *Where there is motion within but not outwards and the total remains unchanged ... we have a parallel in our earth, constant from eternity to pattern and to mass ... there is always water: all the changes of these elements leave unchanged the Principle of the total living thing, our world.*
>
> —PLOTINUS, *SECOND ENNEAD*

The questions of metaphysics are often very, very old, yet like the waters of the earth they are constantly renewed. What happens to a question when it ages? It is asked and asked again, transplanted from native to foreign soil, shaped into new linguistic forms and saturated with the thinking of countless individuals. We could imagine a raindrop, which begins with a speck of dust that falls towards the earth, collecting greater mass and speed even as its form constantly changes.

But perhaps a raindrop is not the best metaphor for the history of metaphysical contemplation. Philosophy constantly expands, but the moments of its history do not amalgamate into an undifferentiated body. Philosophy gains mass as it moves through time, but it also gains substance. Perhaps metaphysics is more a hailstone, which is tossed by wind and pulled by gravity into cycles of collecting and freezing, collecting and freezing, until it falls to earth with each layer of ice preserved. When we contemplate the waters of metaphysics, we do not simply study an idle object; we encapsulate the oldest thinking within our newer knowledge. We lend mass, substance, and force to what were already its strongest points.

Our passage from Plotinus speaks to this legacy, as well as our role in it. The possible shapes, sizes, and layering of hailstones are infinite, yet each is a form of the fundamental substance, water. Like the hailstone, being itself constantly accretes around its most central aspect (what Plotinus called the monad, the One or the Unity), emanating outwards from the center and yet, paradoxically, making up the substance of being itself in the process.

Substance has been the primary object of study in ontology since Aristotle. Substance, like the water that makes up the hailstone, is essential Being, the ultimate uniformity, but also the ultimate heterogeneity. In this conception, being emanates from itself, and must thereby combine with anything it encounters or touches. The being of water is its impurity—it is absolutely solvent, absolutely able to admix with every other element. "Pure" water, like "pure" being, would mean absolute death, a wiping clean of all the diversity of life. And thankfully so, for, just as it is the universal impurity of water that allows for life, so too it is the impurity, heterogeneity, and universal solubility of Being that makes metaphysics itself both possible, and necessary.

Image: Eamon McMahon, *Thunderstorm, Northern Saskatchewan*, 2008

Water Scores *Carolyn Turner*

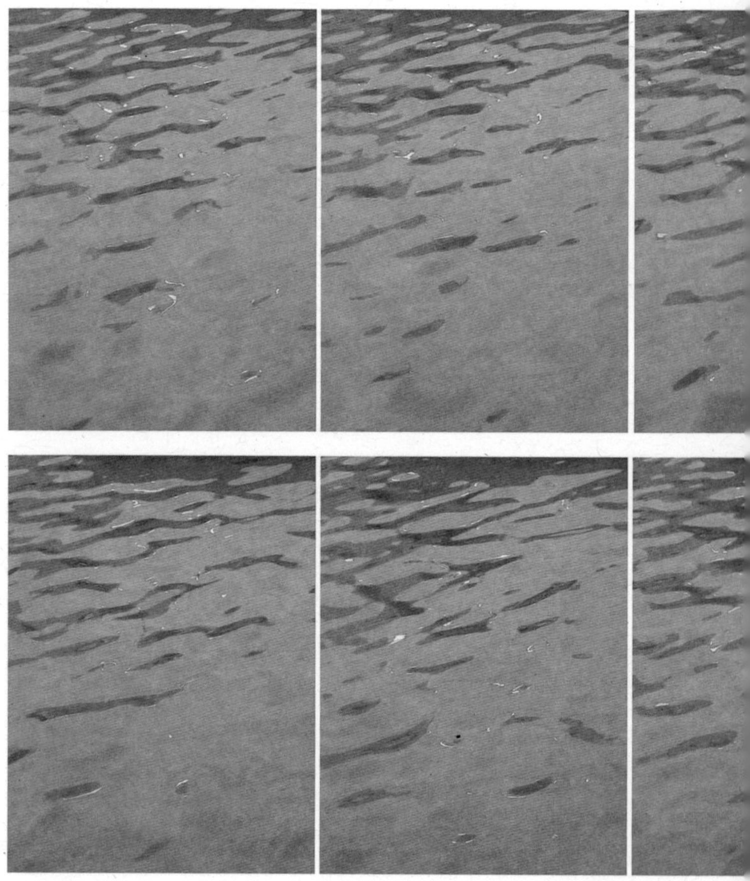

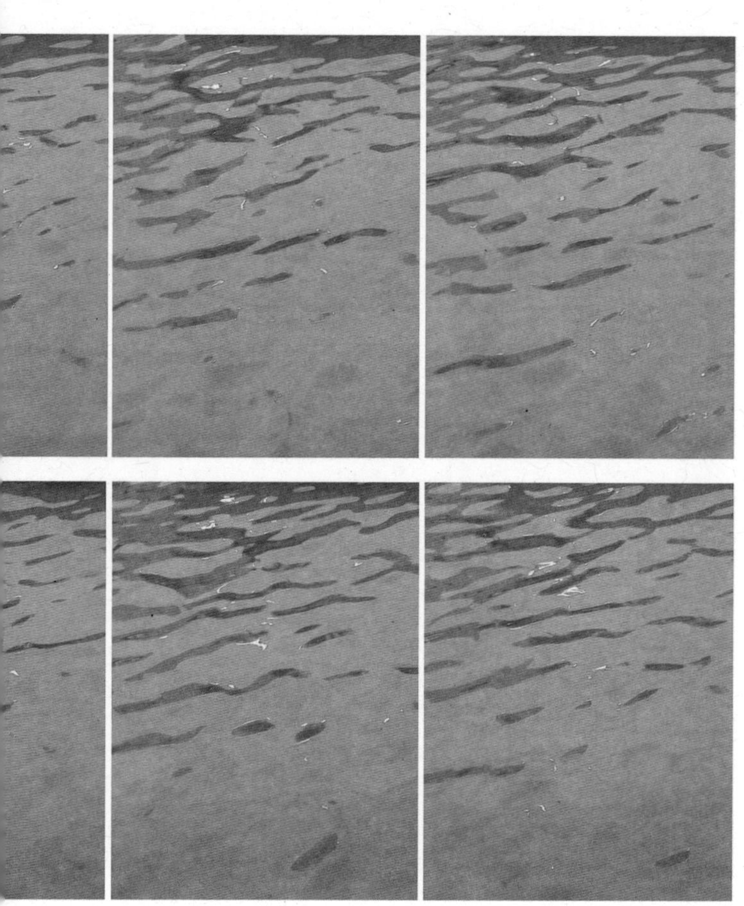

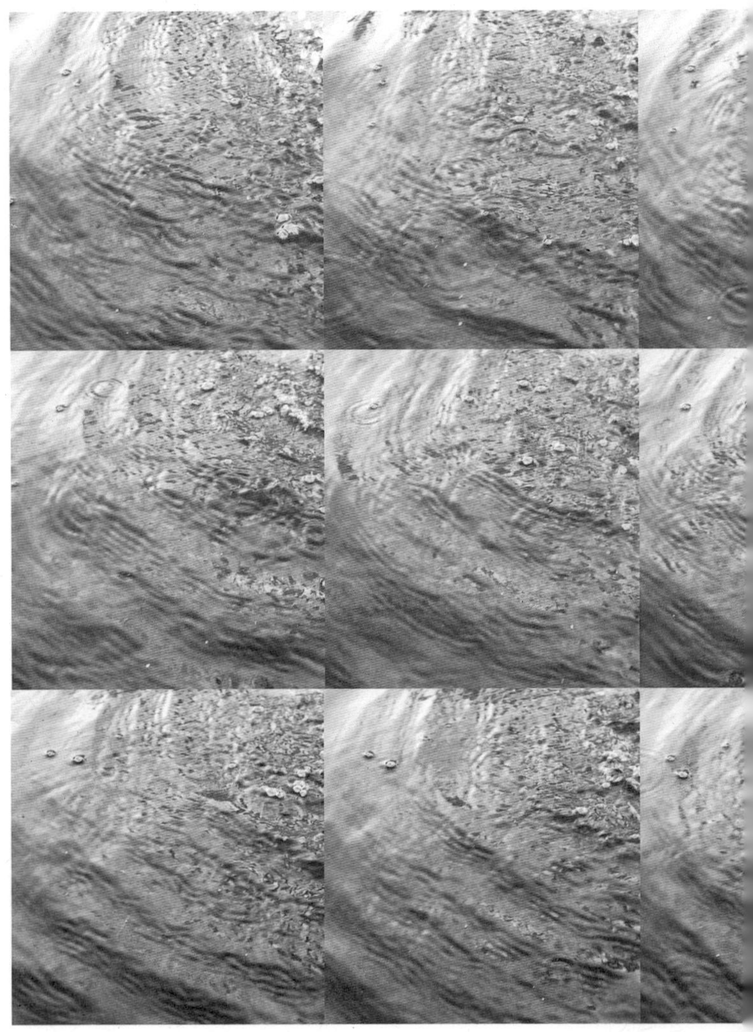

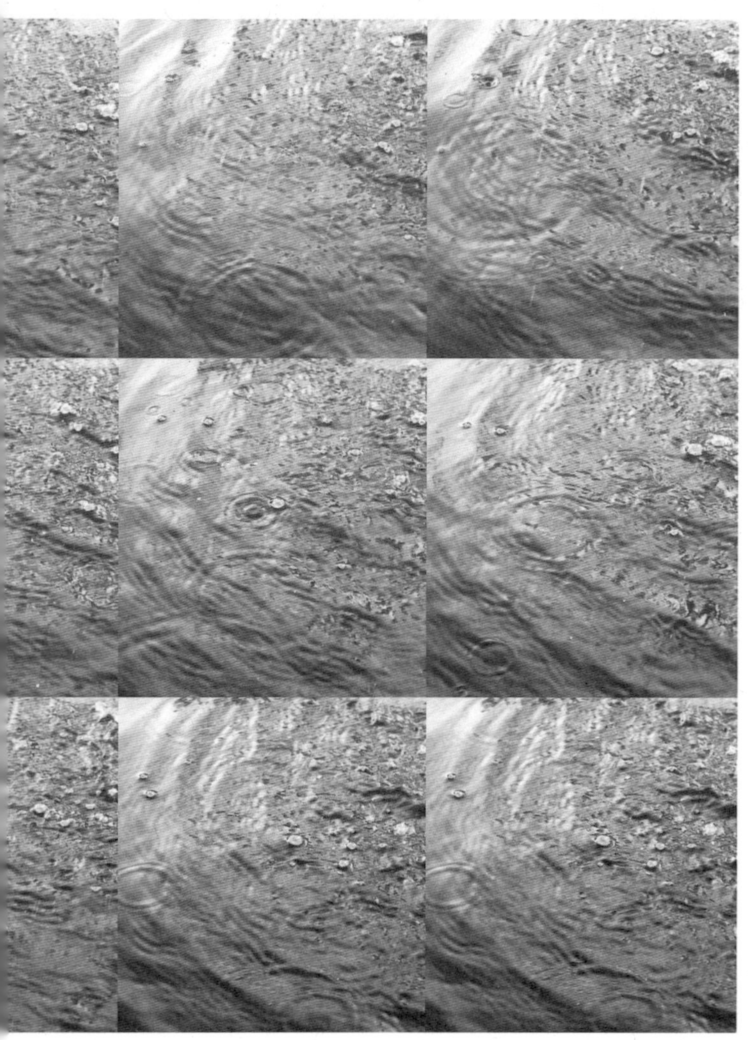

Star Land

There has existed a family of whales that have lived deep in the ocean for many thousands of years. In the recent past, the small fish that the whales eat had become scarce and some of the whales would explore further near the coast to look for food. One night a young whale swam up the coast and into the mouth of a river. He came upon many lights and a star-like moon that seemed to move across the ocean and back. Maybe there is food over here, he thought, and he ventured closer to the stars that seemed to hover in mountains on the land. As he swam up the river the water became darker and harder for him to see and he stopped to catch his breath on the surface. A gleam of light on his back caught the eye of an eagle and she flew down to a nearby log. "The water is so dark here," said the whale, "and I've come so close to the stars. I don't know how to get home." "Those aren't the real stars," said the eagle. "Set your sight higher to see the real sky. Then you will know your way home. And if you hear the sound of the metal boats behind you, keep going. If you feel the oil that sticks to your back, keep going." The young whale looked up over the bright lights and saw the real moon and the northern star and swam in the direction he knew to be home. As he struggled through a current that resonated deep into the ocean, he watched a large metal fish create huge waves in the water. He had become very tired and hungry and as he looked around he could see the star land receding. Finally, as the sun was beginning its path over the ocean, he could see his family of whales surface on the horizon and the eagle fly up to the first warmth of the sun.

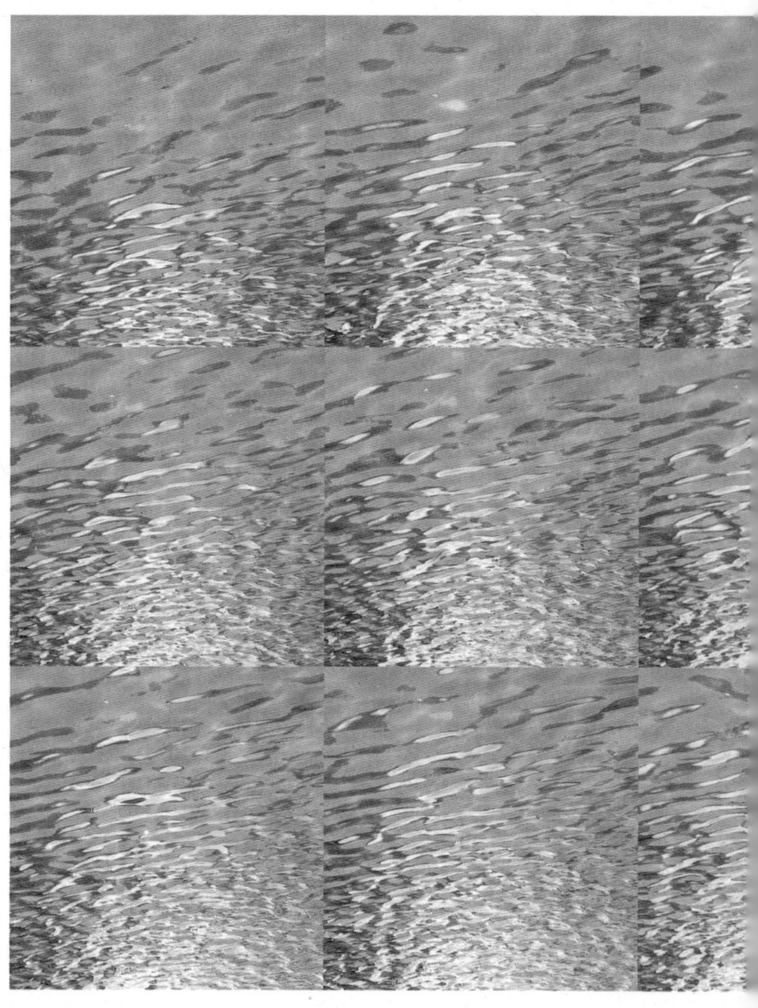

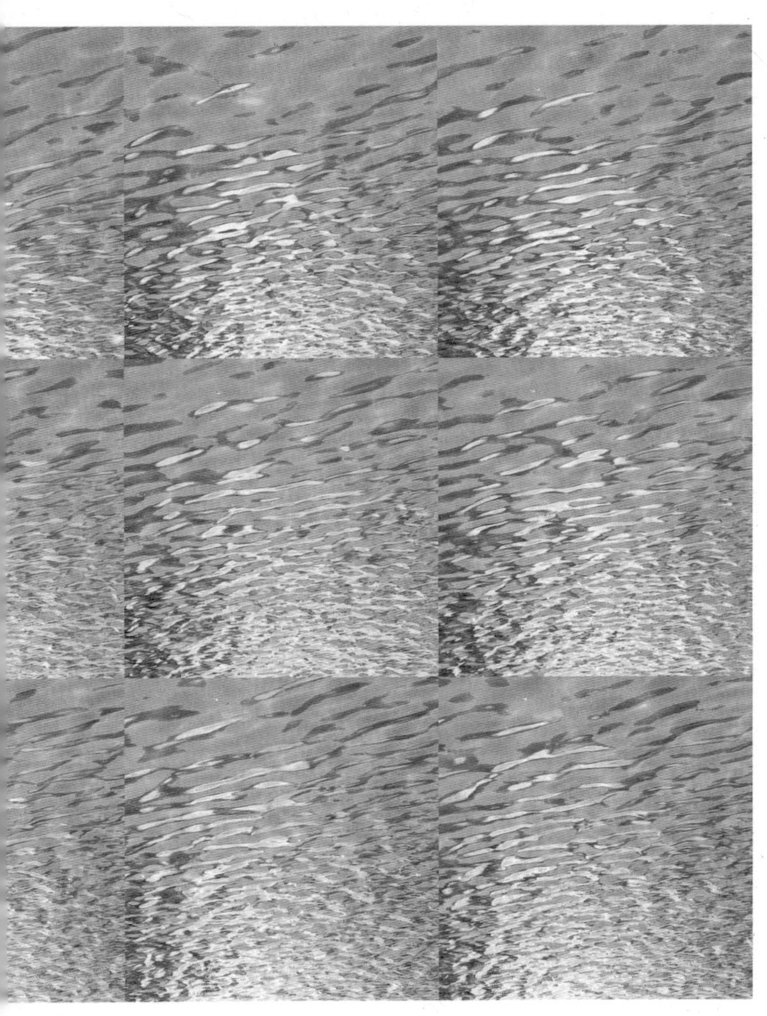

Images from "Water Scores" Series © Carolyn Turner 2008. "Star Land" © Carolyn Turner 2009.

Watching Water *Meredith Carruthers and Susannah Wesley* The sublime majesty of Niagara Falls has inspired artists and poets for hundreds of years, from the first documented encounters to the present. Visitors have responded to the force and power of the falls with feats of daring, endurance, heroism, and even magic. Fictional, rumored, or real, these acts have shaped the perception of Niagara in our cultural imagination.

In the 1840s and '50s the German-born artist Godfrey Frankenstein (1820–1873) painted Niagara Falls over and over, from a hundred angles, under all seasonal conditions. His work verged on obsession. In nine years of observation, sketching, and painting, he produced more than two hundred studies, and his finished work, Frankenstein's Panorama, consisted of approximately one hundred works on canvas. These were mounted in sequence on a display mechanism, which unfurled to provide viewers with a procession of images of the roaring water. The panorama stood 2.4 meters high, and the canvas rolls of paintings were over 300 meters long; it must have been an impressive sight for the visitors who gathered in large numbers to see it in New York City in 1853. Frankenstein kept his audiences' attention with lively narration and music as the images moved seamlessly through a large picture frame on the stage. After crowded performances in New York, Frankenstein's Panorama toured other American cities to great acclaim. According to press notices of the time, the project was received as a wonder in its own right. Unfortunately, very few traces of Frankenstein's panorama have survived, and the paintings themselves seem to be lost completely. The experience and impact of Frankenstein's Panorama must be left to our imagination.

In the spirit of Frankenstein, we created a new panorama of Niagara Falls, a collage of recent images gleaned from photo-sharing websites. Many of the twenty million annual visitors to the falls, out of a desire to remember and share their experience of the thundering cascade, have documented the encounter in photographs, then uploaded the pictures to sites such as Flickr and Facebook. The resulting collective online archive rivals Frankenstein's obsessive waterfall survey, and ranges from snapshot portraits to carefully composed images of the falls sparkling with ice, illuminated by night in rainbow hues, or alive with birds. These photos form a stunning tribute to the enduring enchantment and power of Niagara Falls.

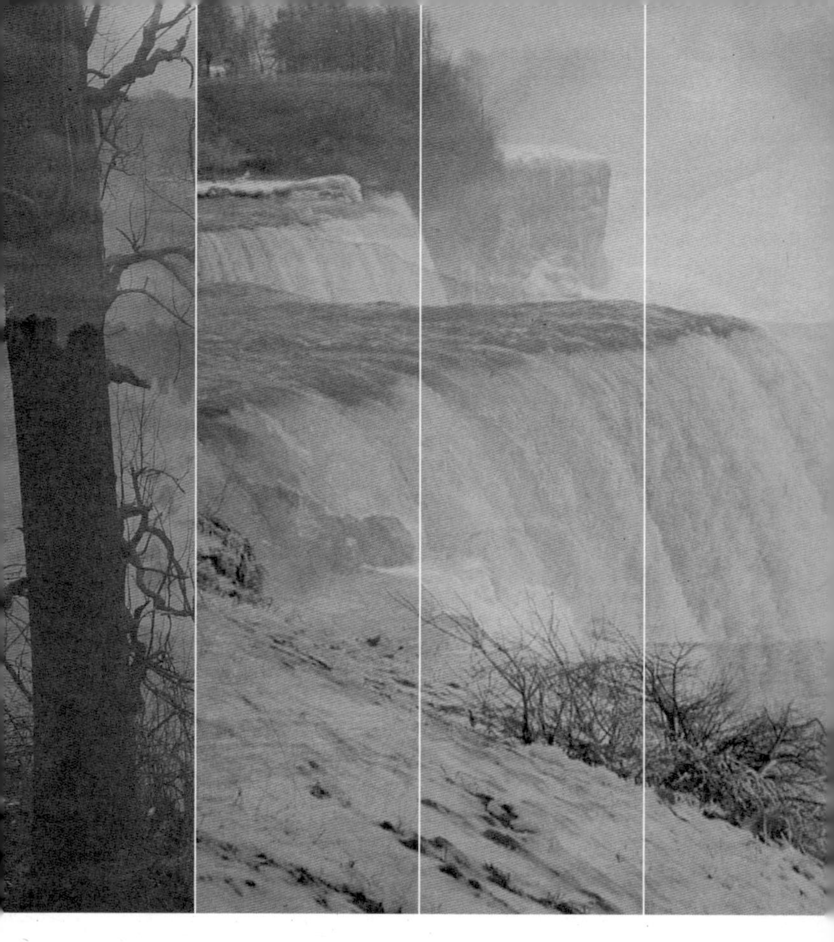

Fresh Green SPRING!
American Bridal Falls. funfuncheese, April 9, 2007
Niagara Falls in spring. Moustache-man, April 12, 2007
American Falls from the *Maid of the Mist*. Niagara Falls. New York. USA, May 2005

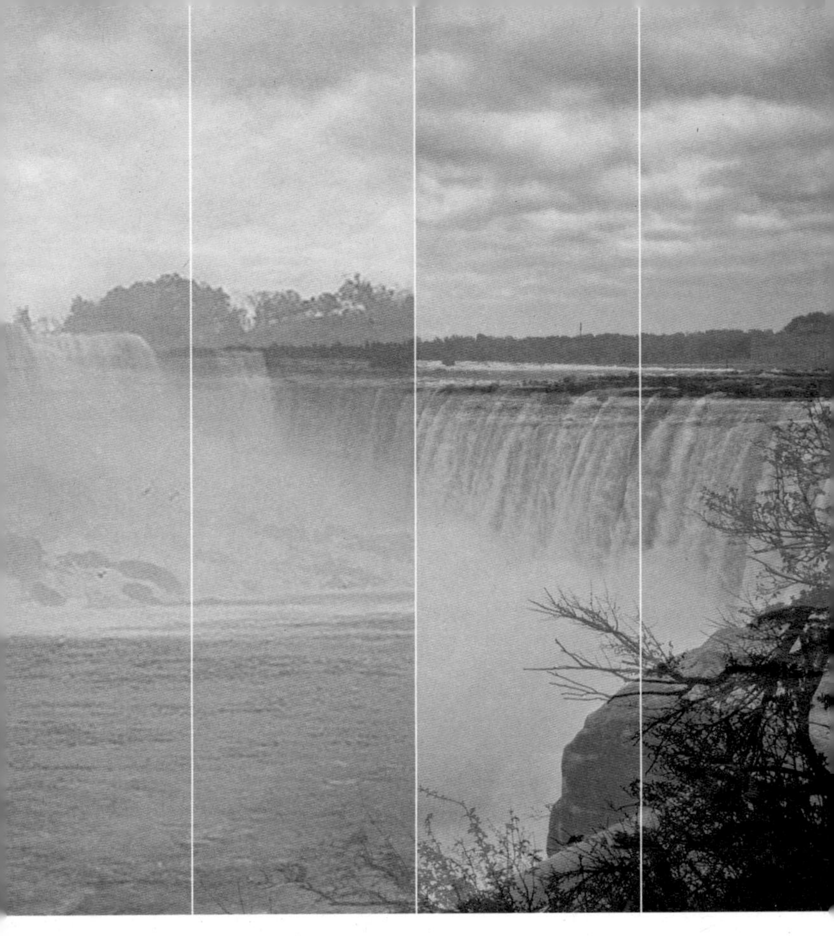

The Horror, May 23, 2005
Horseshoe Falls. jchalmer, May 26, 2008
Horseshoe Falls, Canada—A combination of three pictures of the Horseshoe Falls shrouded in mist. jchalmer, May 29, 2008

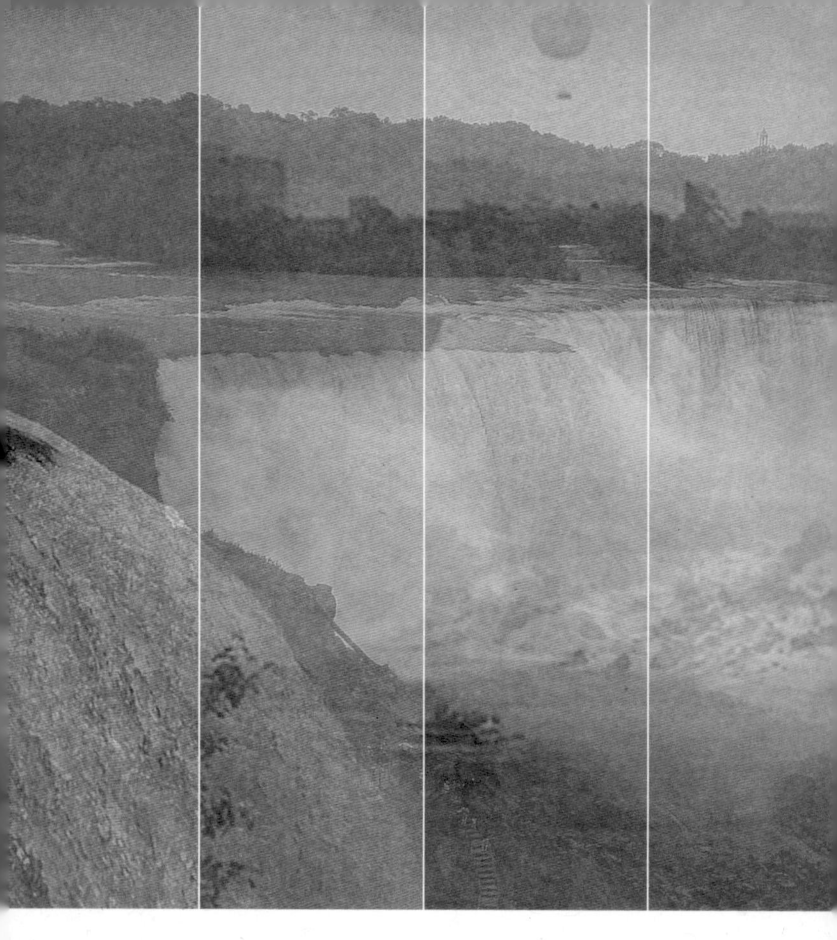

In the heat of SUMMER!
Niagara Falls—American Falls. Wil Zoetekouw, summer of 2003
Photograph by Yangyongwei. August 2, 2005
Niagara Falls. hokorii, August 2, 2007

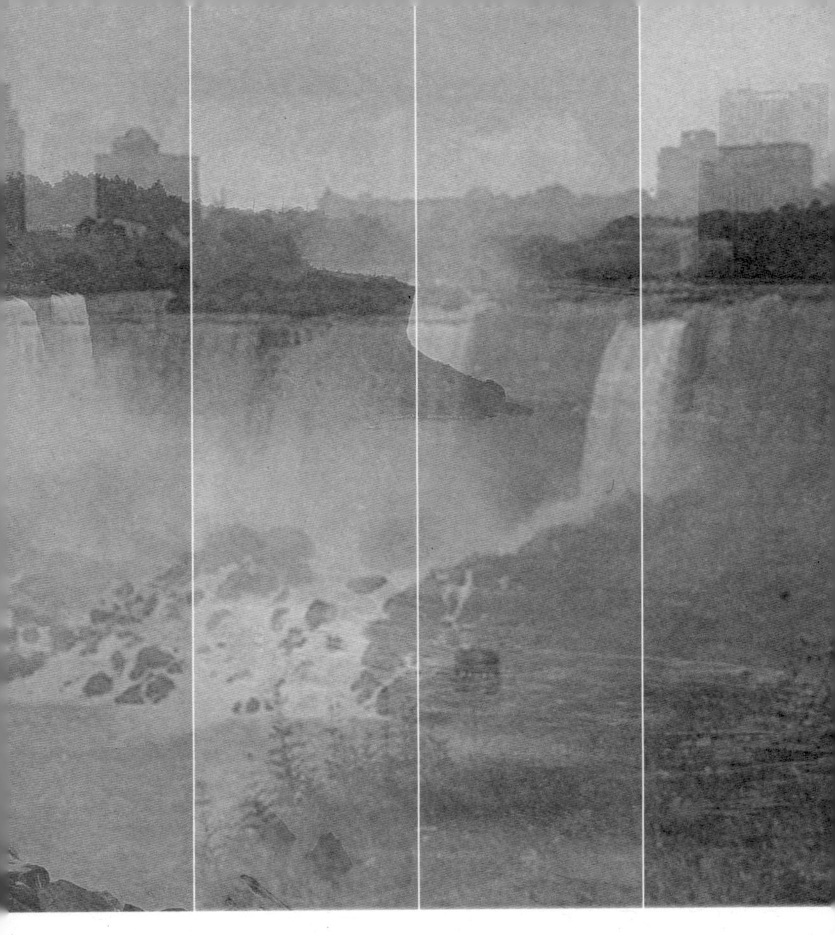

Niagara Falls—Canada. Difo&Natura, August 6, 2008
Horseshoe Falls (Canada). Wolfgang Staudt, August 10, 2006
Niagara Falls (American Falls). digitaleye81, August 19, 2007
The Falls. Dreamer, August 24, 2006

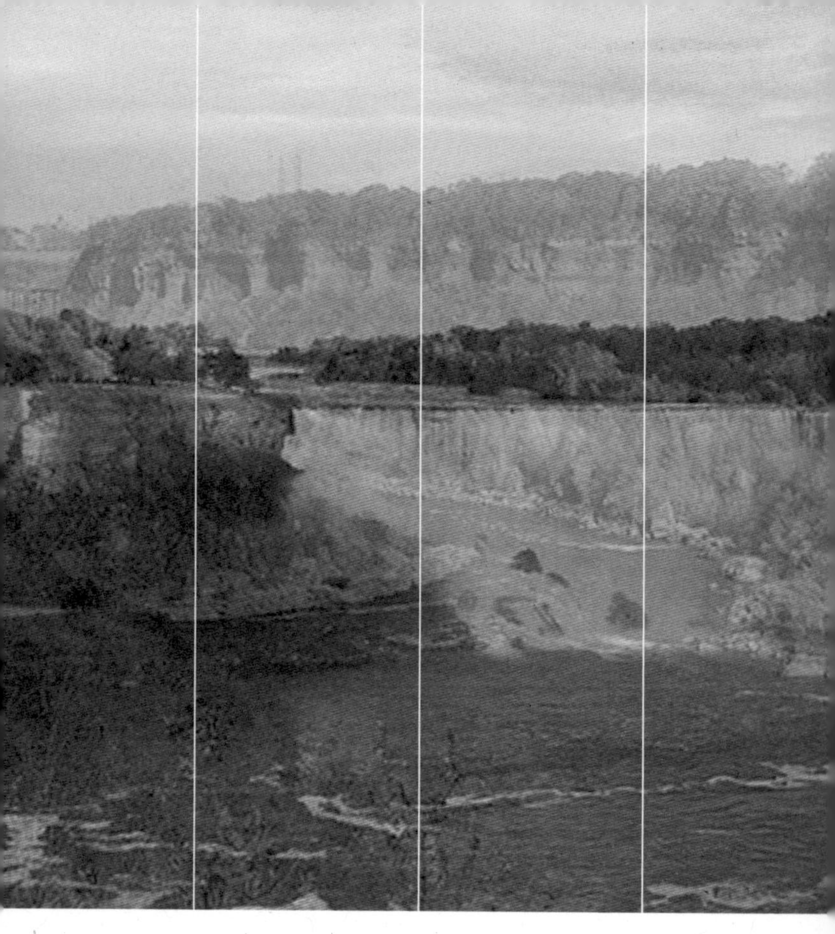

In the Gorgeous Beauty of AUTUMN!
My favorite photo! Autumn approaches. Xtradot, September 8, 2007
06-kan-329. Herrwick, September 27, 2006
Whirlpool Rapids Bridge, Niagara Falls. bridgepix, September 30, 2003
Both sides of Niagara Falls. Josiah Norton4, October 13, 2008
DSF_9496. Crobart, November 2, 2008

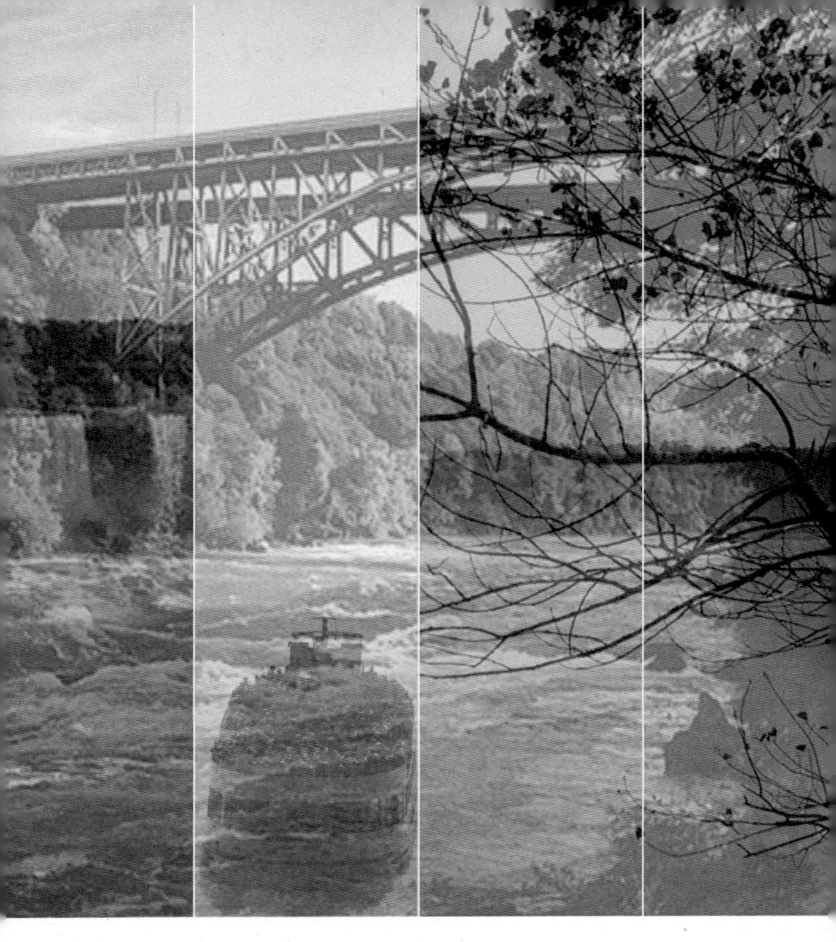

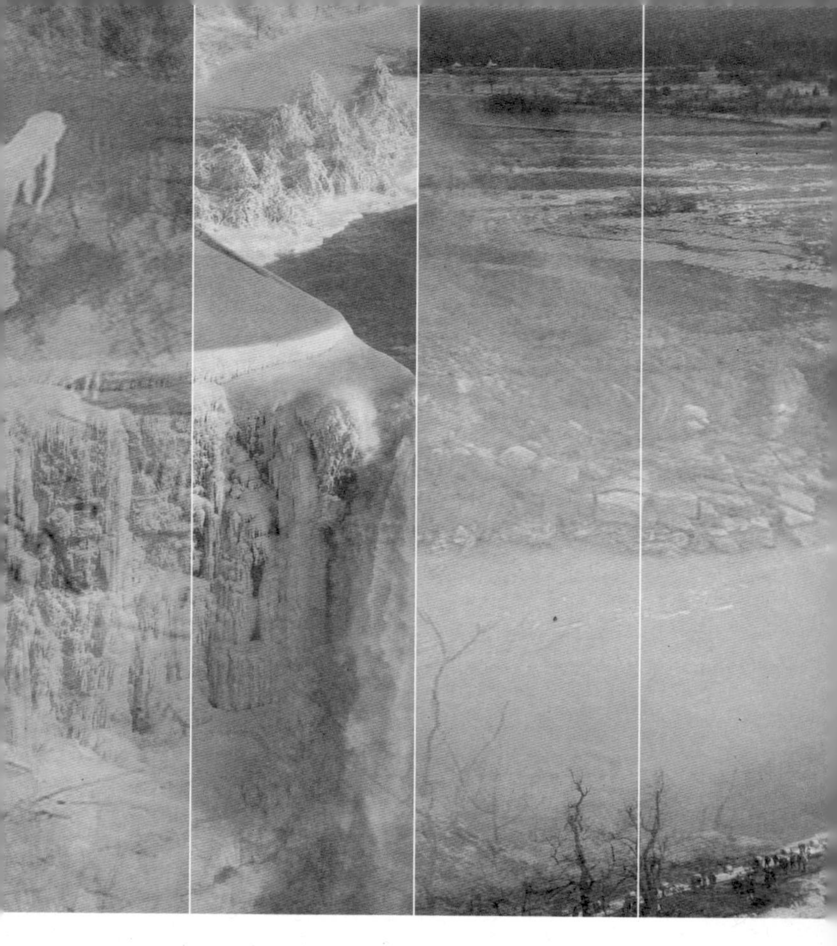

And in the Dazzling Splendor of WINTER!
Horsehoe Falls, by night, in winter, Niagara Falls. elPadawan, January 5, 2008
Niagara Falls winter. AndyPdLA, February 11, 2006
Magnificant (Niagara Falls). flipkeat, February 16, 2008
Left Edge— Niagara Falls. flipkeat, February 16, 2008
Niagara Falls. wstubbs2, March 14, 2006

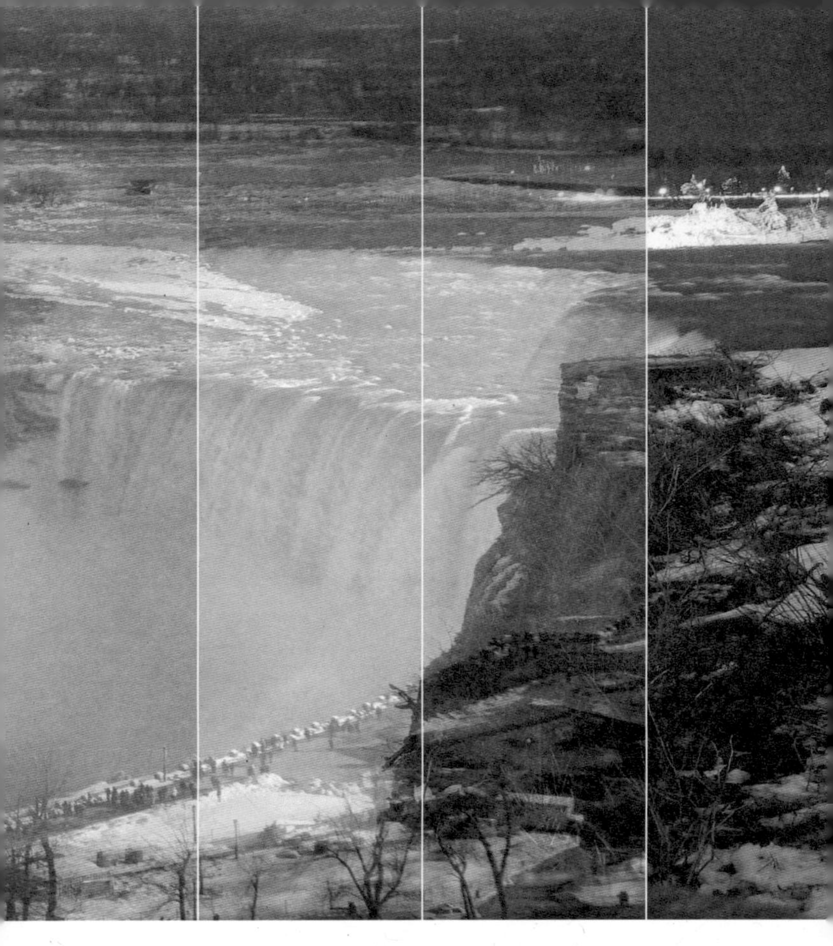

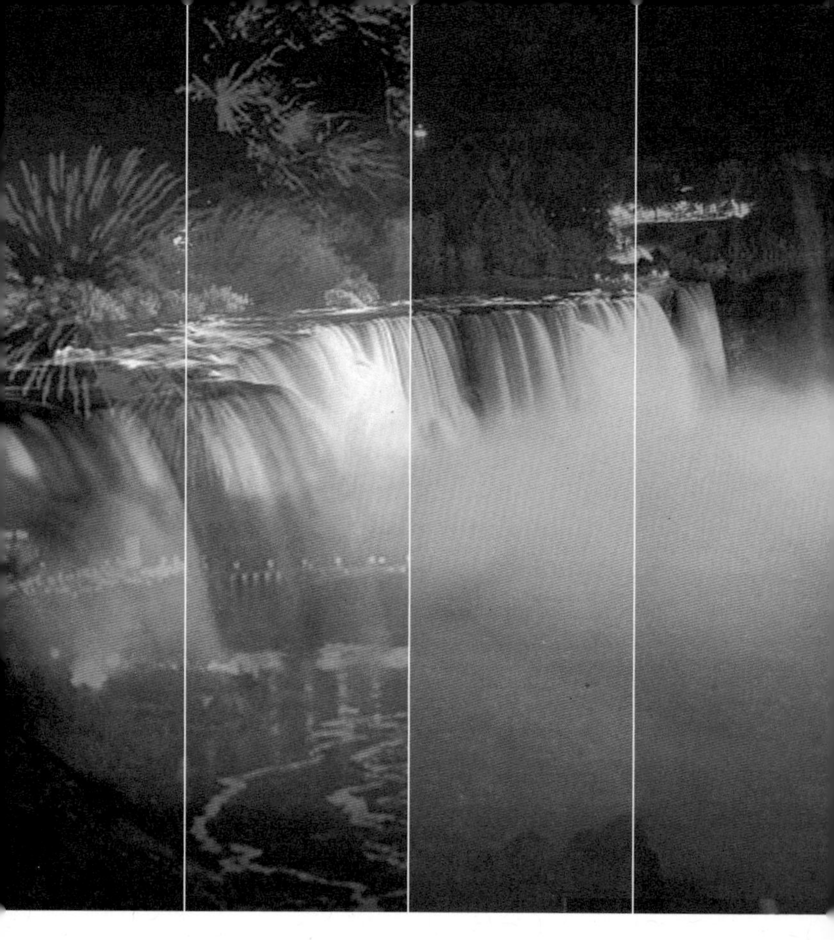

By the MYSTERIOUS AND SOLEMN MOONLIGHT; by the ILLUMINATION OF FIRE!
Niagara at night. Jeremy Cliff, August 5, 2007
Fireworks at Niagara Falls. Skootie, July 20, 2007
Niagara Falls at night, from the Canadian side, illuminated by colored spotlights.
Booknero, December 3, 2005

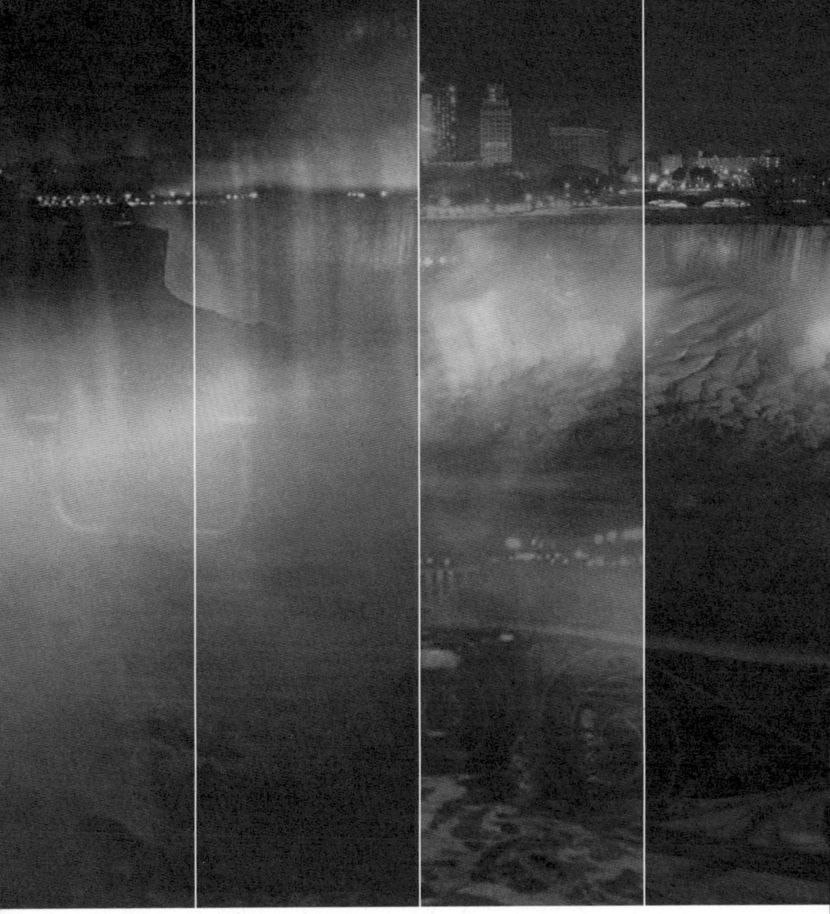

A view of Niagara Falls at night from Canadian side. shahz4, December 26, 2007
American Falls at Niagara. jj8rock, March 17, 2008

Note: The titles of these works are drawn from Frankenstein's enthusiastic proclamations included in the original broadside announcing his Panorama.

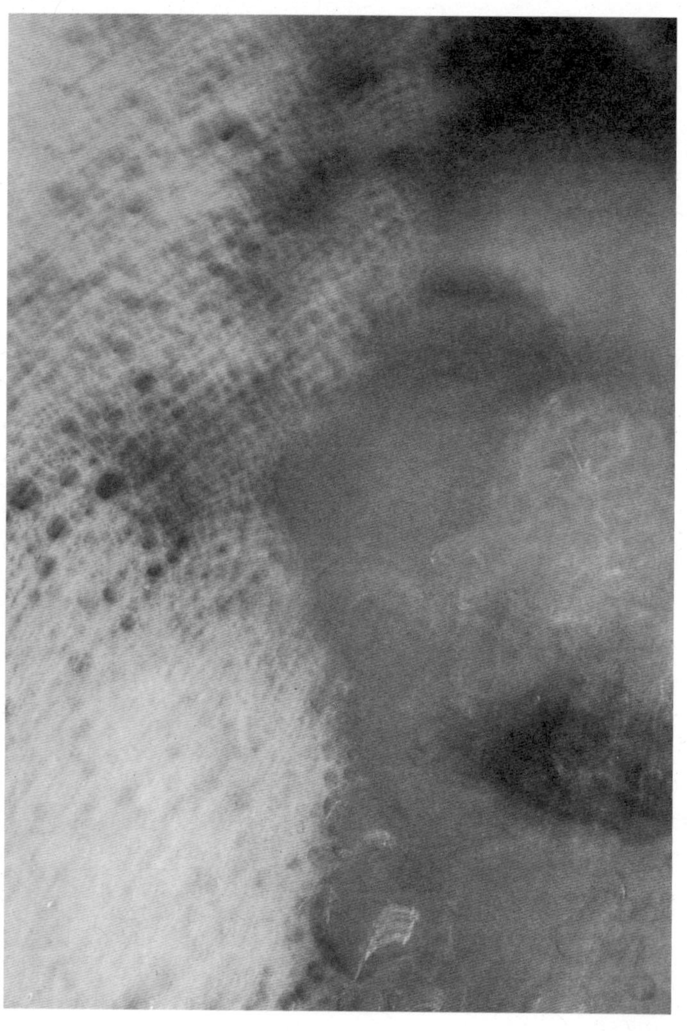

Women, Water *Jowita Bydlowska* There is an archetypal image of a tragic woman in my head. She is drunk, with black running down her face, her body in the bathtub. She is sitting up, her knees are sticking out of the soup of bubbles, a shaky cigarette going out from getting wet in her hand. This bath is the last remaining comfort, a fleeting one—the water is cooling. But she can't move—she is stuck here.

2

The painter Pierre Bonnard painted his dark-haired wife, Marthe, bathing many times. In one painting Marthe is lying with her chin resting on her chest; her elongated body is square-angled, her legs seem to be pushing against the edges of the tub.

Her eyes are closed. Her mouth is tiny, quiet, focused. She's floating.

Rumour had it that Marthe was depressed. She could lie in the bathtub for hours (the water turning cold). Hours: her skin is peeling off, slowly in a cloudy, pale mist.

Pierre Bonnard married Marthe twenty-four days before his long-time lover, Renée, drowned herself in a bathtub. Renée drowned herself because Pierre refused to leave Marthe for her. Renée was also a subject of Pierre Bonnard's paintings.

Pierre Bonnard painted Marthe's first bathing portrait on the day of Renée's suicide.

In an attempt to hide from Marthe the fact of Renée's presence in his paintings, Pierre Bonnard repainted Rénée's hair golden blonde.

3

As a child, I was one of the two creatures that used the bathtub in my house. The grownups were keen on taking showers. To me showers were too predictable, practical, like dressing warmly for winter or eating vegetables for your health.

I loved taking baths. Anything could happen: A fishing expedition, a water battle, a race between my wound-up toy frogs, slow sinking of islands made out of knees.

I discovered then that while on your back, if you dove under and exhaled through your nose, the water and the bubbles above you made enough noise to block any possible outside noise (loud voices).

One day, one of the grownups tore into the bathroom and pulled me out, halfway through my diving trick.

What was wrong with me! Hadn't I heard? There was a child in the family that drowned in a bathtub! It was no joke. That baby drowning. It was true. It only took a little water. It only took a minute. The tub wasn't even full. This was fist-shaken at me.

Bathing became stressful.

After that episode, as soon as the cascading water had ended and there was no sound but the wet clicks of my submerging body, my family's shouting would start. With time, I would turn the water off and yell right away, unasked, I'm alive I'm fine I'm fine! to avoid the drowned child and its story.

Eventually, it was too exhausting to bathe and scream. I graduated to showers and discovered the pleasures of the shower head against various parts of my curious body.

4

When I was a child, the other thing besides me that spent a long time in a bathtub was carp. Every Christmas my father would bring home a live carp and it would be released into a tub full of water until it was time to whack it in the head in the evening, then turn the whole mess into a horrible dish called "carp in jelly." In Poland, it has been documented that the usage of water is the lowest prior to Christmas. This is because all Polish bathtubs are occupied by carp.

Because I had hysterical tendencies, I was told to stay away from the Christmas carp, but of course that only made me want to be around it more. I sat with my chin against the white edge of the tub watching the giant, ugly fish breathe in and out with its dumb mouth, feeling hopeless for both of us.

There was nothing I could do to help it. True, I could pull the carp out of the bathtub and carry it in a plastic bag somewhere (but where, where?!) where I could release it into the water, set it free. There were a number of problems. There was nowhere to go (the nearest manmade pond was frozen or at best filled to the brim with duck poop), it was winter (the carp would freeze in the bag), I was a child (not allowed to cross the street), it was late and we were in a large city (perverts, drunks, priests, etc.).

I never made the attempt to rescue the Christmas carp. I understood the futility of a fantasy. I watched my family members chew the fish and I could not believe I was related to these barbarians. Eventually, I became vegetarian. I stayed that way for thirteen years.

5

But first, I got drunk. It was a New Year's Eve, a week after the last carp's death, my last year in the old country. I was with my friend Anna, outside, looking for trouble. We found it quickly: A bunch of kids from the wrong side of the tracks (kids of: alcoholic parents, divorced parents, parents in jail) with a couple of bottles of cheap, sweet wine. They passed me a bottle and I drank as if I'd never drunk anything in my life before. As if I had been thirsty for the past thirteen years, until now, this moment, when the magical liquid started to fill me with love, light, laughter, and goodness... I drank until there was nothing left.

I came to in a bathtub filled with cold water, back in my old apartment. My friend Anna and one of the kids from outside were staring at me, terrified. "We thought you were dead," said the kid.

"Ohshutup. We were trying to revive you," said Anna, pointing to the tub. "And it worked. You were totally passed out."

I remember shivering and giggling, as they helped me out of the tub.

I remember thinking that I had to do this thing again.

The drinking thing.

6

As an adult and when broken-hearted, I started taking long baths again. I would have some morbid song playing on the stereo, a bottle of wine, water around me and a phone nearby. I would become the archetypal image of the tragic woman in my head.

As I soaked and drank on, I would have an idea to call former boyfriends or my mother. I couldn't call either while sober—too boring and stressful.

I would get white and mushy from crying, hiccuping, sniffling for too long. Getting a (brief) hold of myself, I would dial.

"Where are you?" my mother would ask suspiciously.

"On a fuckin' boat," I'd answer and make some waves. "I have to go now."

Afterwards, I would do like Marthe Bonnard. I would start to fall apart, until all the water got cold. I would stumble out of the bath like a wet, drunk monster, my skin barely hanging on.

The next day, I would have a hangover and a cold.

7

One winter, I almost died.

There were hundreds of different situations that I had created that were sure to lead to my death. Their accumulation should've resulted in one solid death but somehow I kept being here. Stuck.

What is depression? Depression is not waiting. Not waiting to find out what else is out there, why, when, who's around the corner, is anyone. It is not caring if no one cares and if one does care then you do not anyway.

When the worst of depression strikes it is falling to the bottom of a dark lake, slowly, majestically, silently. It is a drowned body expanding on the bottom of that lake into soft flesh, then softer still, until it becomes fish food, until the carp come swimming by and pecking at its fingertips and open mouth.

One evening in the bath, particularly sad (we broke up over drinking and another woman, too) I dove under. I lay on the bottom of the bathtub. I let my arms float, my hair moved above my face like weed. I opened my eyes. My hair danced. The water stung. I shut my eyes. But I remained on the bottom.

I realized then that I was wrong. I was waiting. I was waiting for something after all.

This confused me. Did it mean that I wasn't serious about what I was thinking of doing? There wasn't much time left. Some parts of me were already getting alarmed, asking to be pushed to the surface or drowned; my lungs were tightening into pounding fists. So. Was I serious?

I didn't know any more. I didn't have another profound revelation, no one's arms fished me out, there was no God voice in my head.

But. It just seemed that I wasn't done after all.

How do you fight when you have no fight left?

When it comes to water, you can conquer water with water.

When I was under, with my depression, weakened by tears and then lack of them, the only solution I came up with was to bide my time, since I wasn't quite yet ready to drown all the way, lay my head peacefully inside an abandoned lobster trap, fill my mouth and ears with algae.

I needed to be moving. Otherwise I was going to end up like Renée. Or Marthe.

8

Because I was unemployed, undateable, and drunk all the time, I was often alone and had a lot of time to myself. It stretched and opened before me endlessly, a surface so wide that it crushed me with possibility. But I wasn't waiting for anything so there was no point of looking at the horizon. Nothing was going to happen. I had no expectations. It occurred to me that in order to pass the time—till I had more courage?—I had to do something to simulate movement.

I got a membership at the local pool.

I went swimming every day that winter. I would wake up filled with wet sand from drinking the night before. I would put on my bathing suit and some clothes and march toward the pool building. I was freezing cold by the time I got there and the pool water would be a welcome seeping tea around me as I made my first dip in. And then I swam.

First slowly, without any sort of a system—ten laps, fifteen, backward stroke, break, stretch, whatever—and any sense of the time passing; later methodically, like my childhood wound-up frog, still in the slow lane, making my rounds behind wobbly grandmothers and uneven pregnant women.

Later still—with time measured in quarters and my mind focusing and clearing over my stretching muscles—I went faster. I could feel my arms elongating into decisive strokes, legs flopping with more regularity as time went on, head bravely springing above my shoulders to catch an even breath.

I switched to the moderate speed lane. I didn't know it then but this was when I started swimming toward recovery. In the moderate speed lane, there were more swimmers, an experienced bunch that went round and round, clockwise, steadily. I liked their rhythm.

At night, I went out and screwed men that picked me up or that I picked up at places around my neighborhood. My heart was still broken and I tried to fuck myself into healing. But it wasn't happening. I remained depressed.

And every morning, I would go back to the pool. I swam angrily, my bruised thighs snapping open/shut as I went forward.

I thought of the man from the night before. I thought of my former lover fucking another woman.

If the pain got to be too much, I would dive under and scream. The water muffled everything.

With time, I got tired of being angry. I calmed down. I just swam.

9

I am at a pool again as I write the ending to this story. The water in this pool is the essence of blue; it is soothing and cool, a retreat from the punishing Caribbean sun. I swam a bit earlier, slowly, without counting laps. I caught up with myself a long time ago, at that pool during that deathly winter, somehow.

At this pool, in this moment, I am surrounded by glistening tanning bodies, mostly women, some without their tops on. I have my top off as well and I keep one hand on my stomach in that universal gesture of a pregnant woman. Inside me, the baby is suspended in my own water. We are still but not stuck.

Opening Image: *Under*, Jowita Bydlowska (2008)

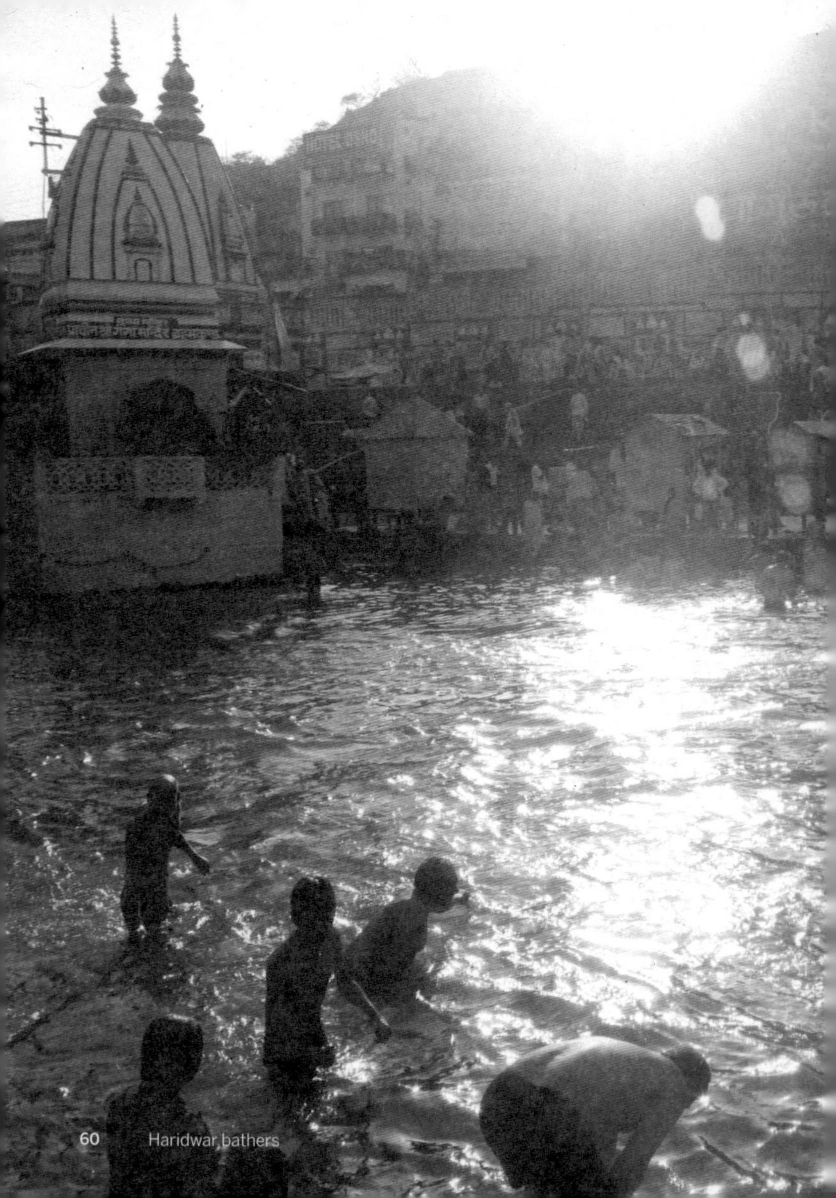

Haridwar bathers

The Public Bath and the City *Christie Pearson*

...it is something within me which knows very well that it would be useless for me to take pains to appear strong and display self mastery, for my wild nature will always be visible through a thousand and one cracks.

—JEAN GENET, *THE MIRACLE OF THE ROSE*

Imagining the city as an enclave of constructed civility always entails a contrasting backdrop of wild nature. But even within urban boundaries, elements of nature are protected through ritual practices, and in spaces where natural forces can have their pleasurable and restorative effects in safe confines. The public bath in its many forms around the world has provided a place for these rituals. Here the distinction between "culture" and "nature" blurs. The public bath as a civic amenity is in decline worldwide, and its value as a ritual has been fractured in the West. At the intersection point of many apparent oppositions—good and bad, clean and dirty, sacred and profane—the meaning of the public bath as an image or an institution reflects our shifting definitions of civil culture, functioning as a mirror of city itself. Now may be the moment to reclaim urban public bathing culture.

Traditions of collective bathing are inseparable from the history of human cultures, with examples stretching back at least to Mohenjo Daro, a city that flourished around 2000 BC in the Indus Valley of what is today Pakistan. Baths may be retreats placed outside a settlement, like the North American sweat lodge, Japanese onsen, European hot springs, or Scandinavian cottage sauna. Some forms may require pilgrimage to a sacred landscape: Hindu worshippers travel to the Ganges to pray and immerse themselves in its cold waters, and to confluence sites for the Kumbh Mela events. When the public bath moves into the city, it may frame natural landscapes, as in an indoor wave pool or a Japanese open-air rotenburo, or amplify the elemental

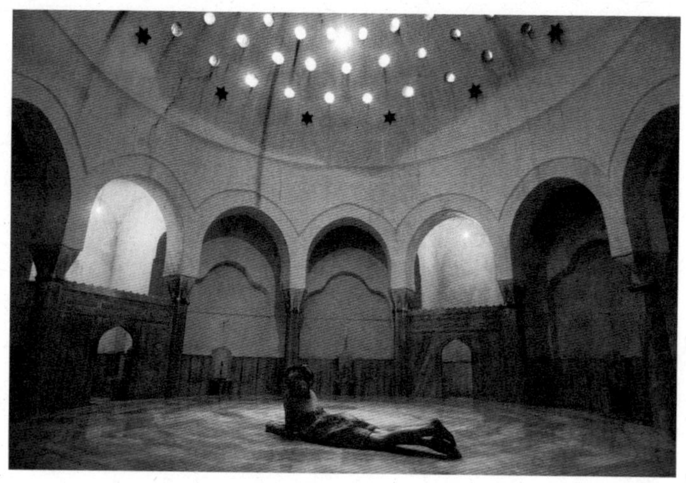

qualities that buildings usually try to moderate, such as heat and cold, wet and dry, darkness and light. These urban bath cultures, like the Roman thermae and balnae, Turkish hammam, Japanese sento, and Russian banya, have formed a cornerstone of their cities' spatial forms and their citizens' daily rituals. They rely on boundaries and rules, which continually evolve as long as the tradition is alive. Etiquette prescribing degrees of nudity shifts: bathers in a disreputable medieval English "stew" were fully naked, while in the eighteenth-century European spa they wore muslin bathing dress. Rules about contact between the sexes shift: when Commodore Perry arrived in Japan in 1853 and the disapproving American eye beheld the mixed bathing in Japanese cities, he pushed the Japanese to more strictly enforce segregation. This taboo took hold eventually—but only in the city. The fact that all still bathe together at many rustic onsen illustrates how flexible the inner and outer behavioral codes must be at the public bath, particularly as the context becomes more or less urban.

 Çemberlitas Bath, Istanbul. Photo Pascal Meunier.

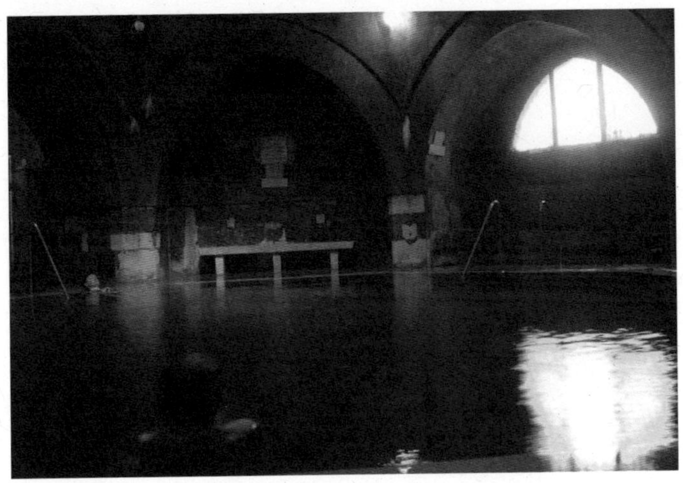

People go to public baths for beautification, hygiene, healing, socializing, amusement, sex, and worship. A visitor to a public bath in the West today may be apprehensive. What are the rules? We fear that our personal boundaries will be compromised and that our public personae will be lost with our clothing. Moral uncertainty always seems to accompany the public bath. Austere Romans echoed Spartan and Greek warnings that a warm bath would make men effeminate. Bathing was held to be an agent of moral decline by early Christians (focusing on outer cleansing could detract from inner purity), and lauded by industrial-era social reformers as a key to moral improvement. Bathing the poor and the ill was considered an act of charity by medieval Christians and Buddhists alike, yet the fear that too much enjoyment would lead to moral deterioration haunts the Western traditions of the bath.

The roots of public bathing lie in the history of the sacred landscape, and many early bathing sites appear to have first been places of

Kiraly Bath, Budapest

worship. Originally chosen for their spectacular and evocative geographies, these sites have the potential to inspire awe before a force greater than ourselves, and even transform us through a meaningful encounter. The urban public bath attempts to capture the power of a wild landscape and its water, and to bring about a catharsis in the bather through its rituals. The rituals that rule the public bath are designed to encourage one to linger and ultimately lose one's sense of purpose, allowing a peaceful space of silence to emerge.

> *The history of mankind goes from natural cave to the artificial cave, from the underground cave to the aboveground cave...*
>
> —NORMAN BROWN, *LOVE'S BODY*

In the West, the act of entering a discreet realm within the city and disrobing with others can symbolize a return to nature, to paradise, or to the amniotic bath from which we are born. The bath in the city may act as a kind of container for a socially constructed version of nature: once inside, we are paradoxically freed to act "naturally" through ritualized rules. Its accepted status as a space governed by ritual makes possible behaviors that the rest of public life rejects. The urban bath has policing agents of various kinds, including your neighbor who will be sure to tell you what you are doing wrong. The many kinds of pleasure that may be enjoyed at the baths are continuously kept in check, adjusting to shifting sensibilities, so that we may approach the powers of the wild in safety, and in a well-rehearsed ritual moderated by evolving architectural and behavioral forms.

The evolution of the Western European public bathing tradition attained its cultural peak, in terms of sheer impact, with the great thermae of Imperial Rome. Modelled on the less luxurious Greek men's gymnasium with its cold showers, these institutions grew to accommodate thousands from many walks of life. At the height of the empire, leisured Romans spent every afternoon at the baths:

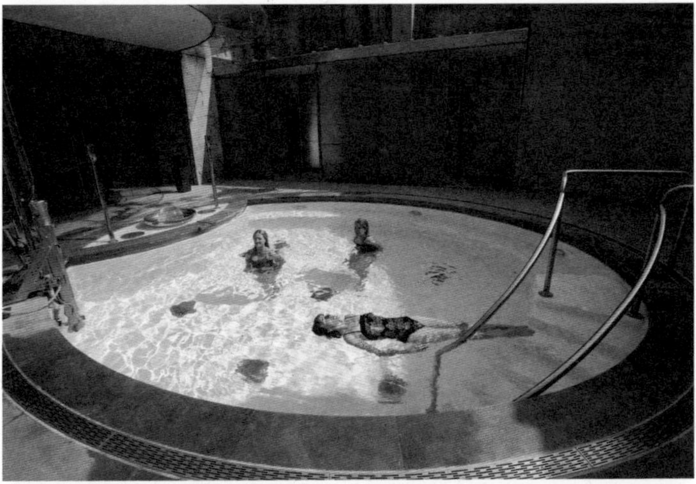

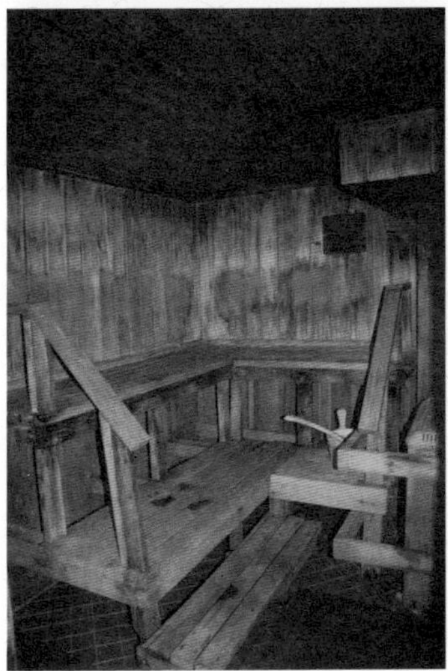

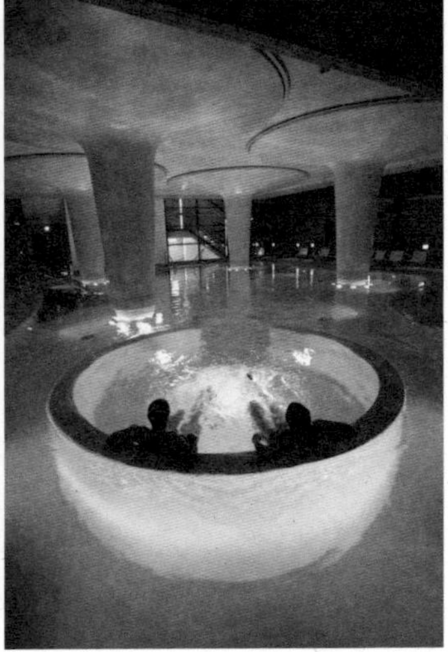

At top: Thermae Bath Spa, cross pool; main pool (photo Matt Cardy); Sauna interior

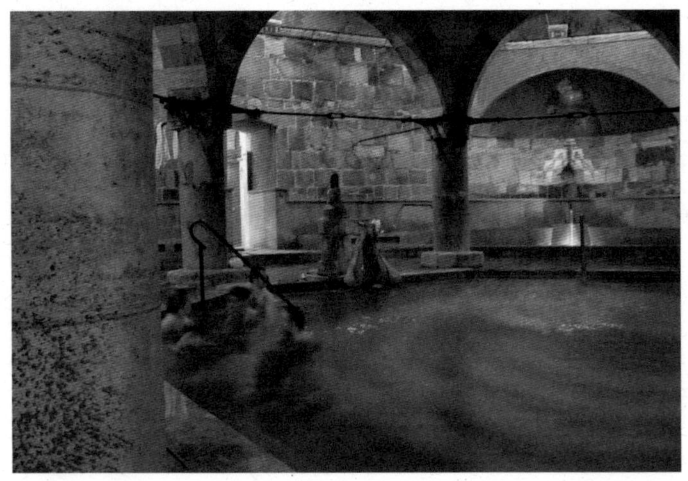

beginning in the dressing room, apodyterium; then moving to a warm room with a heated floor, the tepidarium, to begin a sweat; then the hotter calidarium, which usually contained a hot water source; finishing with a cold plunge bath in the frigidarium. You might decide to warm up by working out in the palestra, or with some poetry reading in a library exedra within the thick wall of service spaces, stores, libraries, and galleries. Vendors roamed, selling food, drinks, depilation, and sexual services. Some emperors tried to enforce gender segregation to prevent the baths being used for sexual activities, but this was never a popular rule: they mostly remained mixed-gender. Roman bath architecture pioneered experiments with complex concrete domes and curves that would later be adopted by Christian churches to evoke another spirituality. A sectional cut through Bath, England, illustrates a history shared by many Western European cities. Beneath the newly developed swimming and health establishment of Bath Spa lie the remnants of Georgian and Victorian spas for the well-heeled,

 Çemberlitas Bath, Istanbul. Photo Pascal Meunier.

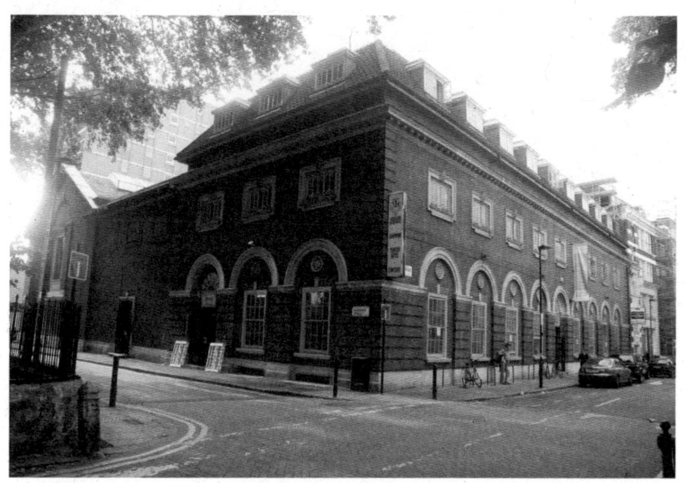

seventeenth-century curative baths frequented by royalty, a medieval monastery where monks used the waters to heal the sick, and the Roman bathing and temple precinct of Sulis-Minerva, all built over the sacred spring of the Celtic goddess Sulis. The civic and sacred functions overlap and continually reframe the hot spring's water.

After the fall of the Roman Empire, public baths and, in fact, bathing of any kind all but disappeared. Medieval bath-houses had to contend with outbreaks of plague and venereal disease and the warnings of doctors, who now considered bathing itself to be dangerous, theorizing that immersion in water compromised the protective covering of one's pores. The monks who had believed in the curative powers of Bath's spring later condemned the public baths for their twenty-four-hour revelry and debauchery, as a rise in illegitimate births caused churches throughout Europe to enforce segregated bathing. Then, when the bubonic plague devastated the continent in the fourteenth century, public baths were closed, abandoned, or openly turned into brothels—only to be cleaned up once again in the sixteenth century when new sexually transmitted diseases appeared.

With the emergence of germ theory in the mid-nineteenth century, opinion again shifted. Cities were crowded and growing quickly as the Industrial Revolution gained strength, and a fear of contagion between classes took hold. Planners came under pressure to provide for public standards of hygiene. England led the way, with measures such as the Public Baths and Wash Houses Act of 1846. American social reformers wishing to "elevate" their citizens responded with the "City Beautiful" movement of the 1890s, when great parks and monumental bath houses were built in cities like New York, Boston, Philadelphia, and Baltimore. These cities quickly had to institute regulations prohibiting showers longer than ten minutes, as pleasure was distracting from purpose. New York children told to leave the summer swimming baths on the Hudson would dirty themselves to be allowed re-entry. People were enjoying themselves, sidestepping the facility's proper hygienic function. The moralizing approach to public improvement attempted to prevent this slide into decadence, away from the good life of virtue into the good life of pleasure.

What survives today in the modern Western city is a splitting and compartmentalizing of programs that were, for a time, coexistent. Cleansing as a moral duty has given way to scientifically endorsed health programs, including showers at public swimming pools and gymnasiums, or in your own home hygienic unit. Bathing is now also a beautifying treatment at the day spa, and is the scene of eroticism in gay male bath-house culture—a vibrant expression of the public bath incorporating sensuality, sexuality, and cleansing. There are some exceptional places where men and women can bathe together; far fewer where erotic expression between women or between genders is tolerated. While still gender-segregated, a sauna or hot tub at a recreation centre is a rich social space that offers many opportunities for communal conversation or private reflection.

Non-Western traditions of public bathing have changed less over the centuries. The Finnish sauna, Japanese sento, Russian banya and Turkish hammam are examples of living traditions, under siege yet still a vital part of communal life. In these enduring forms of the public bath we glimpse what our societies could have. Public baths in Islamic societies are traditionally situated adjacent to or as part of the mosque complex, which includes community centers and schools. The plan of Istanbul reveals cells of residences and businesses, each served by a central hammam and mosque. The architecture of the hammam is an Islamic riff on or repurposing of Roman baths. Its iconic central form is a cosmic dome, and the baths themselves are a sequence of courts, where the largest space flips you up into a constructed starry night sky. Men and women are allowed at different times, or in different areas. The bather first passes through the camekan, a space for smoking, drinking tea, and lounging, to which you will return at the end. In a warm disrobing room you put on wooden clogs and wrap yourself in a pestemal (Turkish towel). Next is the largest space, the hararet, where you sweat lying on an elevated, heated platform. Either alone or with the help of a tellak (attendant), bathers may receive a massage, soaping, and rinsing at one of the perimeter basins. Although once inside the hammam the sexes are segregated, literature makes clear the historical opportunities for liaisons en route to the baths, underscoring the baths' apparent aversion to excessive purposefulness. In Ottoman Turkey, groups of women would spend entire days soaking and gossiping at the hammam, bringing with them their children, food, drink, and musical instruments. At many points in history, this has been the only public social space accessible to women who were otherwise sequestered in the home.

Bathing friends are the best of friends. —JAPANESE SAYING

Urban recreational spaces, such as public pools, gardens, and parks, periodically create festival conditions by temporarily dismantling social hierarchies, allowing us to choose to be participants or spectators in the collective. The urban public bath is different; it makes detached spectatorship impossible. A gentle saturnalia is evoked when we enter the city outdoor swimming pool in summer, or the sauna at the YMCA. The creative excitement of the festival is present, but here it is restrained, guarded by its ritual use. This public bathing tradition is also invoked when urbanites head to a beach, lido, or hot spring with our friends and family: a foray to the edge of the wilderness. Our identities lose articulation in our bathing costume, and we become free to re-create ourselves. We could make more of these opportunities to remove or loosen our identities.

This is something that public spaces of worship also attempt. As a sacred opening to contemplation, religious buildings and gathering spaces invite collective action, imagination, and play as much as the above civic examples. Overlappings and intertwinings of civic and sacred uses of buildings are well documented; the hammam is part of a civic-religious complex; the church of Santa Maria degli Angeli in Rome is a vestige of the Imperial Baths of Diocletian; medieval cathedrals have housed markets, livestock pens, and dancing. Where rituals of cleansing are explicitly addressed in holy texts of Jews, Hindus, or Muslims, the public bath itself has an overtly sacred role. In Finland, the sauna is a traditional place for giving birth and laying out the dead, and a child is taught to behave here as in church. The Koran admits of only two types of public buildings: the mosque (which contains a variety of social functions) and the hammam, the baths. The idea that the public bath is in some sense a sacred space recurs. It is a secular ritual, at once profoundly personal and shared.

However the public bath is framed, its cultivation allows us to continually renew its role as a free zone of peace and contemplation. Searching for meaning amidst the pleasures of the bath, we find instead our minds silently wandering on the peripheries of perception. Deep reflections are glimpsed through shifting mirror-mazes of light and water. Our private bathtub expands in the public bath, where we find ourselves in our broader home of human community.

 Çemberlitas Bath, Istanbul. Photo Pascal Meunier.

Fossils *Stefan Petranek*

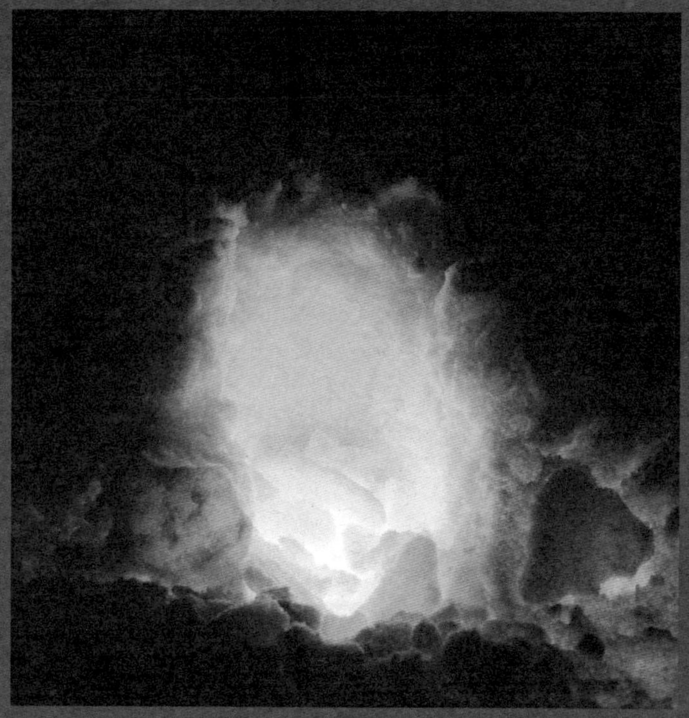

FOSSIL #12

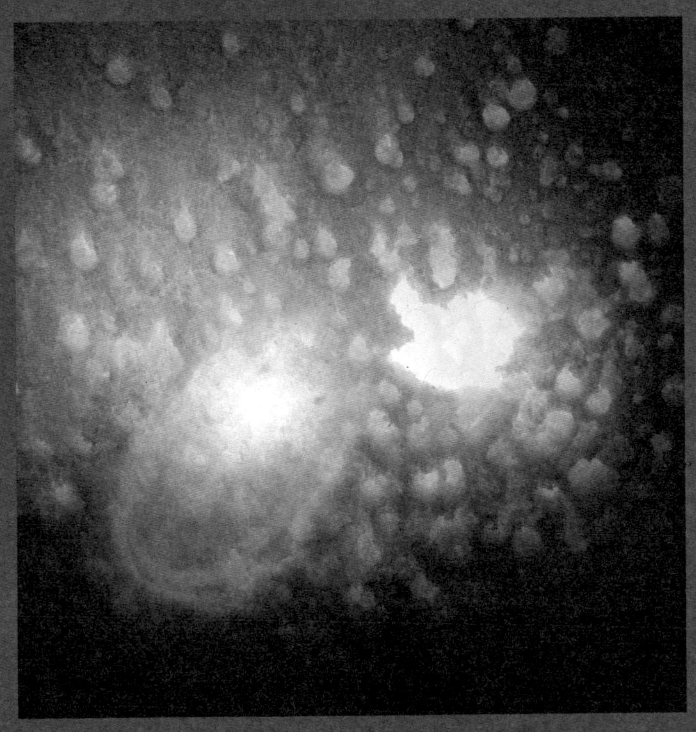

FOSSIL #16

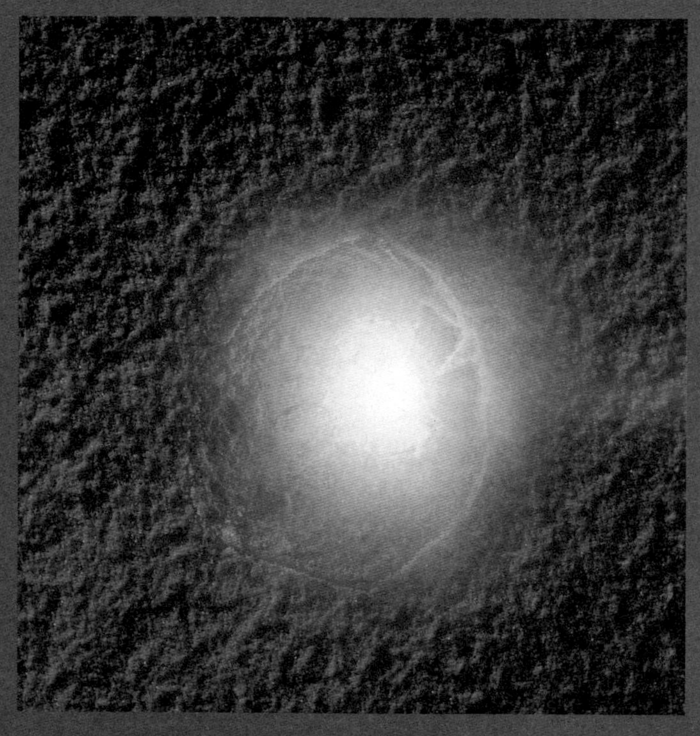

FOSSIL #19

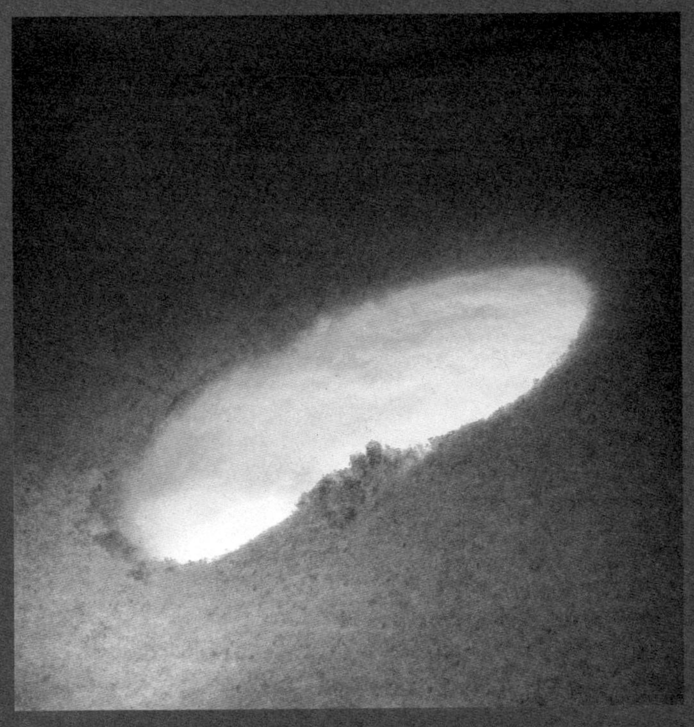

FOSSIL #14

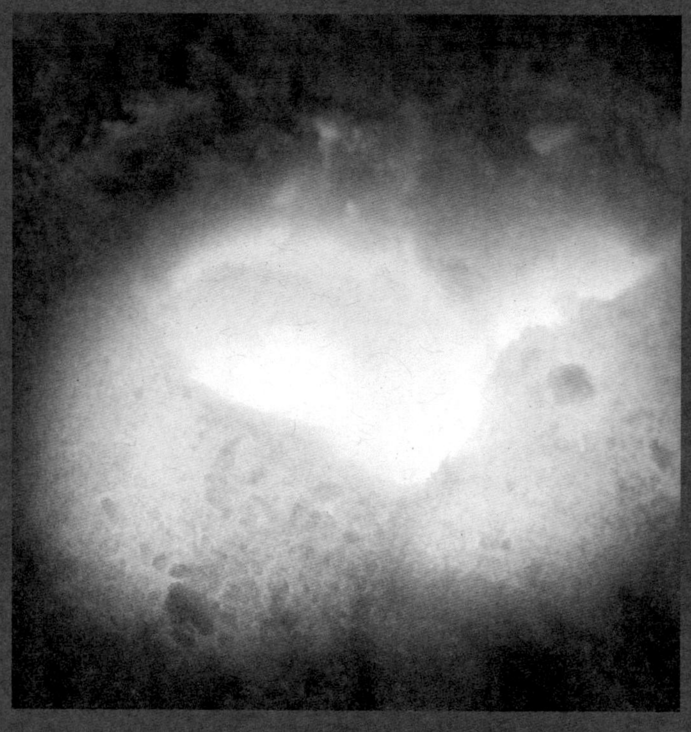

FOSSIL #4

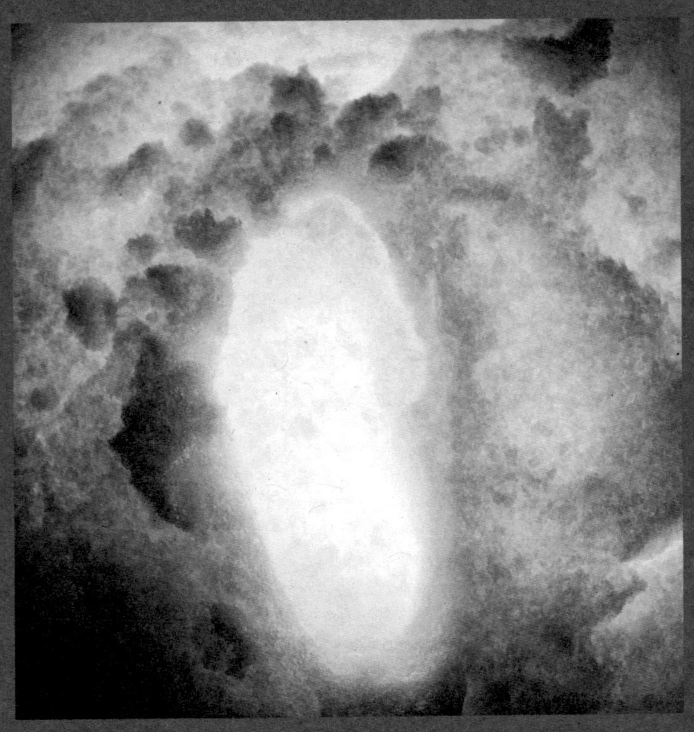

FOSSIL #3

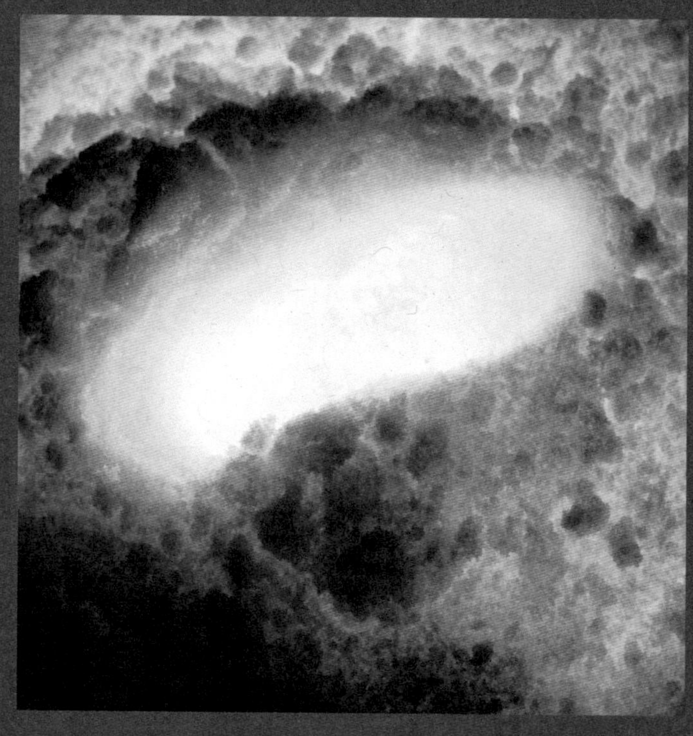

FOSSIL #5

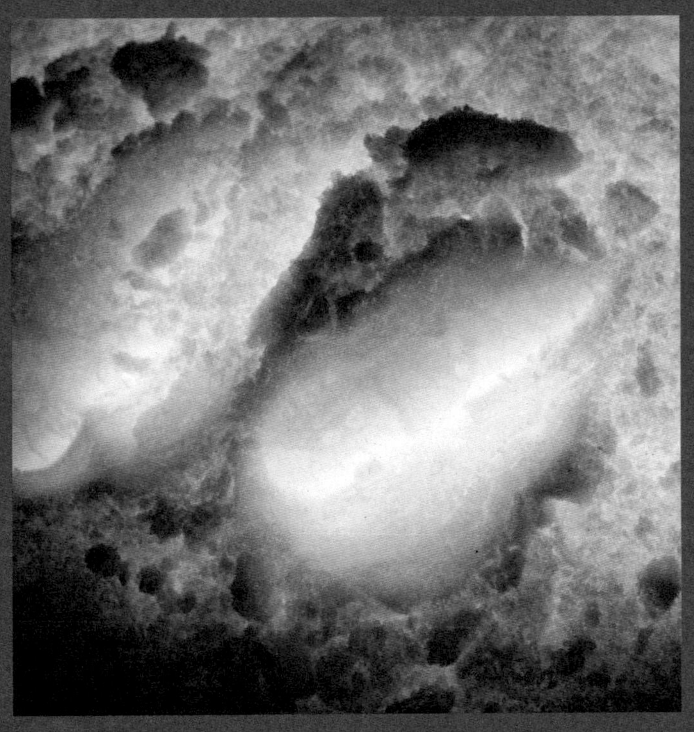

FOSSIL #7

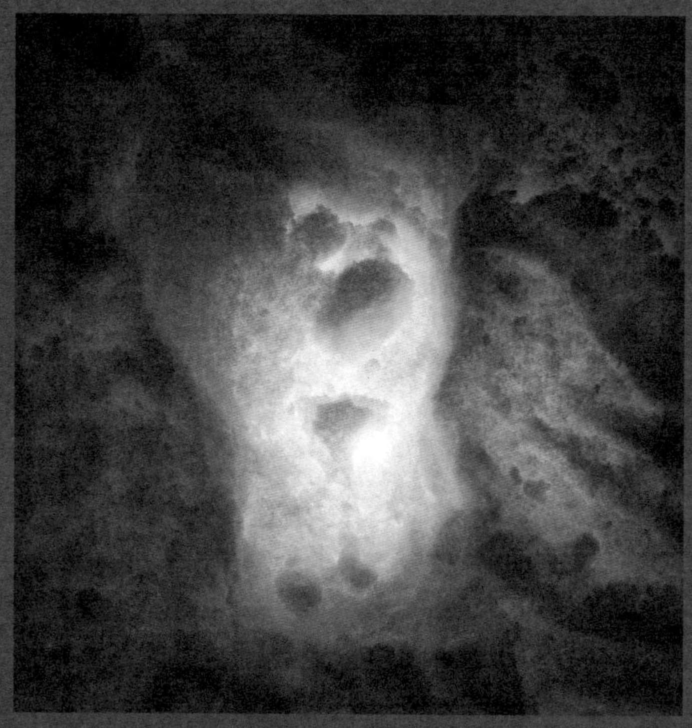

FOSSIL #10B

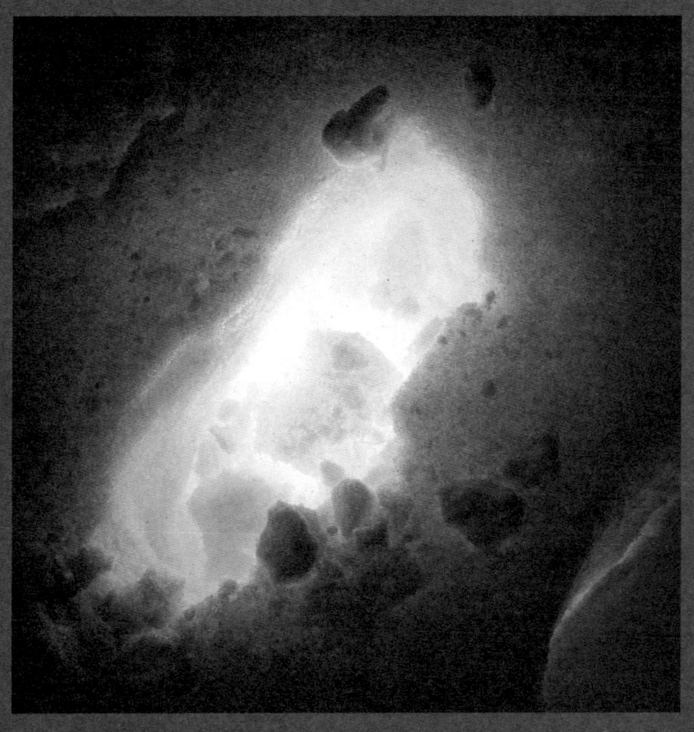

FOSSIL #1

We Are All Bodies of Water *Astrida Neimanis*

I AM THIRSTY.

To define a human being as a body of water is to ask a radical question about what we mean when we call ourselves "bodies."

I PICK UP A GLASS OF WATER

Not so much physiologically speaking, as two-thirds to three-quarters of what we know as our bodies is water. Water is our primary biological necessity—without it we could not move, love, speak, sweat,

AND PREPARE TO TAKE A SIP. WHEN THIS WATER ENTERS ME, IT WILL TRANSFORM ME IN SUBTLE AND SIGNIFICANT WAYS. IT WILL QUENCH MY THIRST, SATIATE MY TISSUES, AND REPLENISH MY BLOOD. THIS WATER WILL SERVE AS VITAL CONDUIT AND COMMUNICATOR BETWEEN THE FARTHEST-FLUNG OUTPOSTS OF MY BODY,

or think. The more radical aspect of the question is that we experience our "bodies of water" as brimming with semiotic potential; their physiological functioning is deeply imbricated in a web of social, political, ethical, and philosophical meaning.

AND PROVIDE THE NECESSARY FLUIDITY FOR ALL OF MY MORPHOGENETIC ENDEAVORS: MOVEMENT, GROWTH, PURGATION. IT WILL ALLOW THE PALMS OF MY HANDS TO BE NERVOUS, MY GENITALS TO BE EAGER, MY EYES TO BE SAD.

Our bodies of water are an inextricable interweaving of both natural matter and cultural meaning, or what Donna Haraway calls *natureculture*.

And this is not in spite of our bodies' wateriness, but rather because of and through it. Water facilitates the stretching and expanding not only of our physical bodies but of our thoughtful, emotional, and imaginative selves as well.

BUT STILL THIS THIRST—

Our watery materiality seeps through the sediments of humanism: bodies as orderly, autonomous subjects. As bodies of water, we are simultaneously subjects and more-than-subjects. In the gurgles of our guts, we might note, entire worlds reside. Our very embodiment comes to be contingent upon excess; our subjectivity floods and overflows. Embarrassing leakages, needy sponginess, visceral transfer stations, and our various other micro-modalities of living all agitate the surface of our seemingly autonomous subjectivity. Gilles Deleuze and Félix Guattari speak of this relation between subjectification and de-subjectification as one between the ***molar***, as a measure of the stability of chemical concentration, and the ***molecular***, as a measure of the unstable collision and reaction of multiple entities. While the molar body is a subject that is sufficiently coherent to take up various human projects, this subject is in turn disrupted by the various subterfuges of the molecular body, thereby revealing other micro-modes of being and being-in-relation.

Consider how moments of intense anxiety, anguish, or terror can result in the sudden and unexpected expulsion of one's water in tears, vomit, urine, and feces. Such eruptions are often beyond the control of our coherent subjectivity. "Excuse the outburst," we say after a tearful breakdown. "How unlike me," we apologize. We treat involuntary flushes of bodily fluids as an embarrassing sign of our brute biology, working against our (totalized, molar) human subjectivity. But water floods the space between the false binary of "metaphysical subject" and "biological life." Indeed, an insufficiency in our bodies of water, expressed through a pressing thirst, can easily disorient

IT BREAKS MY FOCUS, REFRACTING MY INTENTIONS AS EASILY AS WATER WOULD THE LIGHT. MY THROAT SEARCHES FOR SOME FORGOTTEN CACHE OF SALIVA AND MY INCESSANT ATTEMPTS TO SWALLOW DISTRACT ME. I CANNOT CONCENTRATE OR KEEP MY THOUGHTS TRAINED IN ANY DIRECTION. "WHAT DID YOU SAY? WHAT WAS THAT AGAIN?"

and disorganize the molar subject-self. In extreme dehydration, our cognitive understanding of self may recede altogether. Thinking as a body of water suggests that every body expresses the potential to push the boundaries of our conventional sense of bodies as solid, consistent subject-selves.

Deleuze and Guattari are interested not only in the molecular body's disruption of our coherent, subjectified, organized selves, but also in the diffusions and dispersals of our seemingly contained and stable bodies through both space and time. Our molecularity is also about recognizing the transcorporeal assemblages we are constantly entering, flowing from, seeping through. As molecular bodies, we are dispersed, expanding, uncontainable. We live simultaneously as stratified and destratifying, holding on to the shape of a body as a necessary convenience, but also extending out into a constantly reconfiguring web of watery bodies.

AT LAST I DRINK. AS THE WATER PASSES THROUGH MY MOUTH AND RUNS DOWN MY THROAT MY BODY BECOMES

Just as we are mostly composed of water, three-quarters of the planet we inhabit is also wet and blue. Water infiltrates and inhabits the vapor we breathe, the land we till, the multitudinous other species with which we share a watery earth. Even the least watery plant life, eking out its existence under desert conditions, is still made of at least 50 percent water. Add to these bodies meteorological

A HAILSTORM, A CYCLONE,
A FRACTOCUMULUS CLOUD,

and geophysical bodies of water

AN ESTUARY, AN OCEAN,
A BILLABONG, A LAKE OF BAYS,

and we soon recognize ourselves as watery bodies among watery bodies, all sloshing around in a watery world. To understand ourselves as bodies of water thus also asks that we flush our understanding of what it means to be a "body" free of its anthropocentric hang-ups. Our human individuality is thoroughly irrigated. A *body*, as Deleuze and Guattari insist, is any metastable choreography of matter and meaning—something that can hang on to its physical collateral just long enough to be intelligible. As humans, we have no exclusive claim to the term. I am a body of water, as are you, as is a river, a snow squall, a walrus, a tamarind tree.

ALL OF WHICH HAVE BEEN DRAWN FROM OTHER CLOUDS, OTHER OCEANS, AND ARE NOW REDISTILLED AND INCORPORATED INTO THE WATERY MASS OF MY OWN FLESH. ONLY NOW DOES WATER CAPTURE MY FOCUS COMPLETELY. REWINDING THIS WATER'S JOURNEY TO THE GLASS IN MY HAND, I WATCH IT FIRST SLURP BACK UP THE KITCHEN FAUCET, FLOW OUT THROUGH THE INTAKE PIPE, AND SNAKE BACKWARDS BENEATH LAYERS OF FOOTSTEPS, TIRE TRACKS, ASPHALT, AND EARTH UNTIL

IT REACHES THE TREATMENT PLANT ON WOODWARD AVENUE. IT THEN STOPS BRIEFLY IN THE FILTER BASIN TO DIVEST ITSELF OF THE FLUORIDE, CHLORIDE, AND ANHYDROUS AMMONIA IT ACCUMULATED THERE, THEN SWIRLS BACK THROUGH THE SEDIMENTATION TANKS, BACK THROUGH THE RAPID MIX FLOCCULATORS, BACK TO THE MURKY DEPTHS OF HAMILTON HARBOUR. BACKTRACKING FURTHER STILL TO LAKE ONTARIO, THE WATER IN MY GLASS VARIOUSLY EVAPORATES INTO A WARM FRONT, SWIMS BACK UP THE DON RIVER, REVERSES ITSELF INTO A COLD NOVEMBER DRIZZLE. IT BECOMES SPRINGTAIL, BOOKLICE, ZEBRA MUSSEL, LAMPREY.

But what is at stake here is more than a common nomenclature. In our shared ontology as watery bodies we also engage in continuous, reciprocal (yet asymmetrical) exchange. Just as we take our watery being from other (animal, vegetable, geophysical, meteorological) bodies of water on this planet, we in turn pass our water on to yet other watery bodies, some intimately close to us and others distant and dispersed. We embody the hydrological cycle, but this is not a cycle of mere addition and subtraction. Rather, it is a cycle of continuous becoming and transformation. As subsequent watery bodies take on the water we have passed to them, these bodies transpose our watery traces, and translate them anew. As bodies of water, we are all engaged in multiple and multiplied processes of co-creation. In our common wateriness, we are all, as Donna Haraway would say, *companion species*: we, as the vastly heterogeneous watery bodies that inhabit this world, are all continuously caught up in the multidirectional flows of meaning and matter that make us what we are. We are all along for each other's ride; we are molecularly bonded to one another. No body is self-made; we can only come into being through contact and exchange with other bodies—as lodgers, hosts, parasites, guests, and fellow travelers, swept up in the tides of each other's oceans.

Yet still, in each passage of our watery vestiges from watery body to watery body, we express not only a necessary interconnection with other bodies of water, but also an important separation. In our molecularity we still cling to the necessity of molarity, and cultivate in our fuzzy border zones various membranes of differentiation. And so we save ourselves from being swept entirely out to sea.

LATER, AS THIS GLASS OF WATER WORKS ITS
WAY THROUGH MY STOMACH TO MY KIDNEYS,
MY BLADDER, AS I FEEL AN INCREASINGLY
URGENT NEED TO USE THE BATHROOM,
I WONDER WHICH OF MY OWN SMALL MEANINGS
AND MATERIALS WILL BE CARRIED AWAY WITH
IT IN THIS INEVITABLE LEAVE-TAKING:

In these watery intervals of transcorporeality, we are sustained and transformed, at the same time as our difference is safeguarded, by a membrane in between. But at these membranes certain questions fall like single drops from a leaky tap,

SODIUM LAUREL SULPHATE,
DIETHYLALOMINE,
METHYLMERCURY,
POLYCHLORINATED BIPHENYL,
BISPHENOL A,
BUTYLATED HYDROXYTOLUENE,
SULPHUR DIOXIDE,
ETHYLENEDIAMINE
TETRA-ACETIC ACID,

mostly unnoticed, but persistent:

ACETAMINOPHEN,
IBUPROFEN, ESTROGEN,
PROGESTERONE, HALOPERIDOL,
DIAZEPAM, FLUOXETINE...

what responsibility, what response, is called for in these zones of relationality? How might we begin to repay the debts, reciprocate the gifts, relinquish the thefts, of these, our watery bodies?

BUT PERHAPS THIS IS ALSO WHAT I MEAN
WHEN I SAY AT THE SAME TIME THAT I TAKE
LEAVE OF MY EXHAUSTION, IMBALANCE,
JUST AS I MIGHT PASS ON TRACES OF MY OWN
APPREHENSION, RESIGNATION, LOVE?

Finally, then, to understand ourselves primarily as bodies of water is not only a way to rethink the ontology of what it means to be a body, but also to rethink our relationship with the other multitudinous bodies of water with which we are locked in an inescapable relation of *interbeing*. Interbeing, in the first place, is an arrangement of beings, or of bodies. Interbeing

AS MY THIRST RECEDES, I REALIZE NOT ALL
OF THAT WATER IS BECOMING ME. SOME OF IT
PASSES THROUGH MY INTESTINAL WALLS,
TRANSVERSALLY TRANSPORTED TO A NEW
MOLECULAR POTENTIALITY, TO AN INTERIOR
FRONTIER WHERE I AM ALSO FLUVIAL,
WHERE MY VISCERA UNDULATE,
CARRYING YOU DELICATELY UPON THEIR CREST.

is a logic of existence that reminds us that we do not merely live, but that we *live with* and *live through* other bodies, other beings. But this logic of liquid living is not only *of* and *through*—this living is also a living *thanks to* and a living *for*.

YOU, YOU ARE THAT OCEANIC FEELING, WELLING UP IN A SPRING OF AMNIOTIC POSSIBILITY.

It is a living of debt to the other watery beings that have gestated us, from the most distant four-billion-year-old primordial soup, to the most proximate pool of our mother's womb, at the same time as it is a living cultivation and facilitation of the watery beings that will come after us: the zygote feeding off our own amniotic innards, or the earthly humus that in the end we will inevitably irrigate, dust to dust and water to water.

In recognizing that our hydrological cycles are not only evaporation, condensation, and precipitation, but asymmetrical passages and mutual imbrications as well, we understand that ontology is already ethics. When we live as bodies of water, we are compelled to view our embodiment as both relational and consequential, with gale-force power to affect other bodies, in ways both extreme and imperceptible.

MY SOUNDS REVERBERATE
AND REACH YOU ACROSS THESE
INTERIOR WAVES ALTHOUGH I CAN'T
ALWAYS HEAR WHAT YOU ARE SAYING.

Which bodies are becoming more dangerously parched, due to what I pump out of the watery earth? Which are drowned in the floods of swelling seas and warming temperatures, because of what I pump into our watery weather, or because I have neglected the levees? Which are being irrigated only by cholera-ridden water, or mercury-contaminated water, because there are bodies that I choose not to see, not to hear, not to feel responsible for? How might I sustain that thirsty life and help it to flourish?

Some insist that a water ethic must begin with the international recognition of a fundamental human right to water. Indeed, this recognition will provide the potential for redress for some of the human privatizations, commandeerings, and contaminations of our planet's lifeblood. But would guaranteeing our right to water be sufficient to address our more-than-human siphonings, spillages, and containments? An individualistic and anthropocentric human rights paradigm cannot account for all of the bodies I take up and pass on in drips, traces, and vestigial foldings.

Perhaps instead we need to begin by recognizing that we are all part of a radically embodied hydrocommons, and that we must listen to the differentiated needs of the multitude of bodies within it. In this watery web, we are all downstream from one another—as individuals and species, at biological, geological, and semiotic levels—and we all have responsibility for the well-being of the commons. But as bodies both molar and molecular,

THESE DAYS I FEEL TORSION IN MY GUT,
TOES BETWEEN MY RIBS.

we have power not only to contaminate and contain, but to nourish and proliferate as well. We all hold the potential, folded within our own watery flesh, to cultivate our common renewal.

YOU GROW AND GESTATE, BUOYED UP
IN MY OWN PRIVATE BAYOU. AND SO
I WONDER, WHAT OTHER UNIMAGINABLE FEATS
MIGHT A BODY OF WATER BE CAPABLE OF?

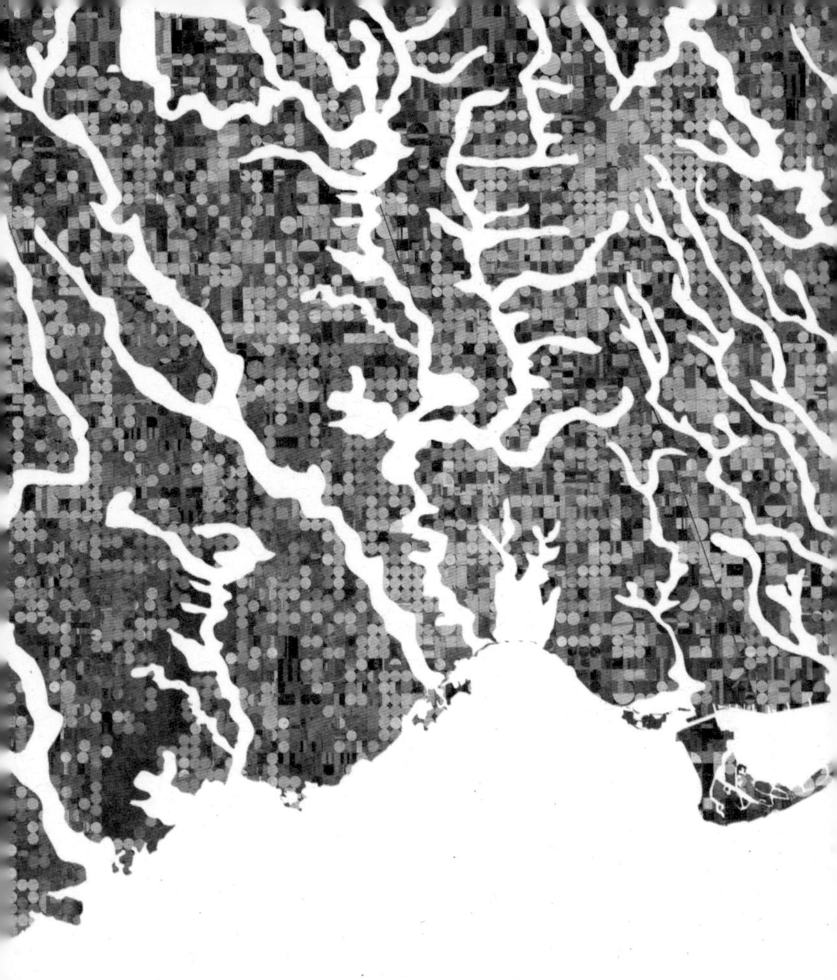

The final form of Ravine City, showing the full restoration of the Toronto ravines cutting through the city fabric.

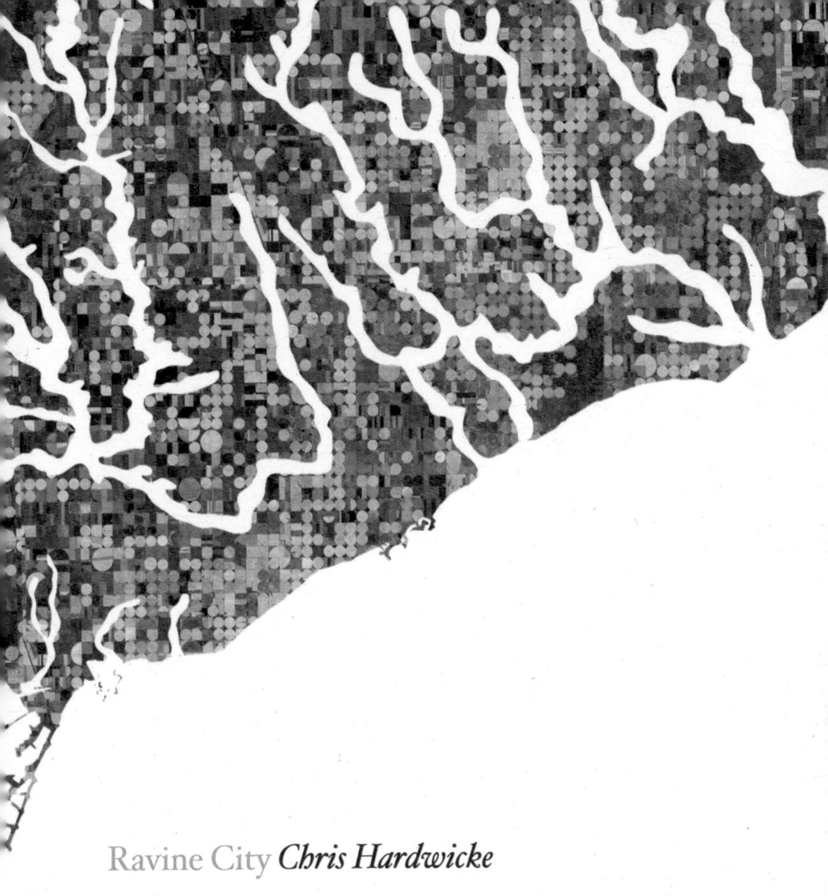

Ravine City *Chris Hardwicke*

It's a city of ravines.... Through these great sunken gardens you can traverse the city beneath the streets, look up to the floating neighborhoods, houses built in the treetops.

—ANNE MICHAELS, *FUGITIVE PIECES*

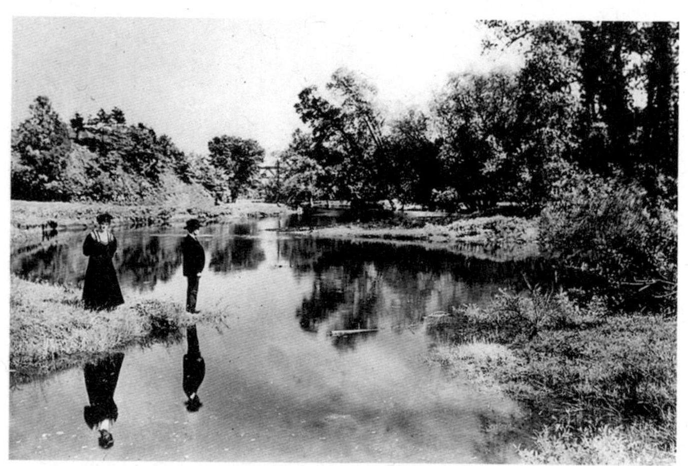

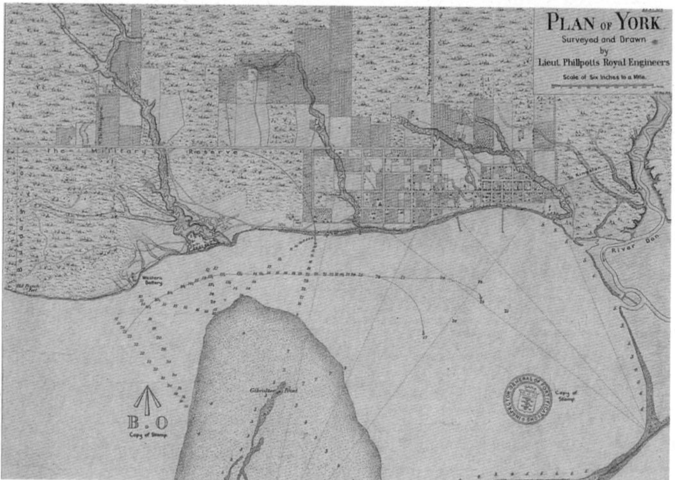

A bucolic Toronto river scene *circa.* 1910; a detail of an 1818 map of Toronto shows the original city grid cutting across the natural ravines

The world's water supply is finite but our water use grows exponentially. In the twentieth century, while the world's population tripled and migrated to cities in great numbers, the consumption of water has grown sixfold. As our population continues to grow and urbanize, so will our demand for water. Given that our cities are designed to impair rather than sustain natural hydrological cycles, these developments will have serious consequences for the environment.

One answer to this condition is Ravine City, a model for a transformative new urban hydrological infrastructure. Using Toronto as an example, Ravine City proposes to unbury the streams and rivers that used to flow through the ravines, and restore the city's natural water cycle. The restored ravines would be lined with terraced housing and gardens that work in symbiosis with the city's watersheds.

In a natural hydrological cycle, water moves around, over, and through the Earth via precipitation, run-off, infiltration, evaporation, and condensation. Most rainfall infiltrates the ground. In large cities like Toronto, water flow has been channeled to control flooding, and most of the natural rain infiltration is prevented by the impervious surfaces of rooftops, roads, and parking lots. Rainwater is collected and routed into the storm sewer system—a vast hidden infrastructure—and as it flows across hard surfaces it picks up heavy metals, oil, grease, chemicals, animal waste, pesticides, bacteria, phosphorus and other pollutants before being discharged into rivers and streams, lakes and oceans.

Instead of closing off the city's hydrology, each building in Ravine City will be an integral part of the ravine ecosystem. The adjoined terraced roof levels of the housing development will follow the top edge of the ravines, forming a continuous connected growing system. This artificial ravine will function much like the natural ravines: controlling water flow and regeneration, as well as cleaning the air and creating habitat and biomass. Maintained and operated by the city, this new topographic infrastructure will connect to the natural ravine system and operate as a second level of public open space.

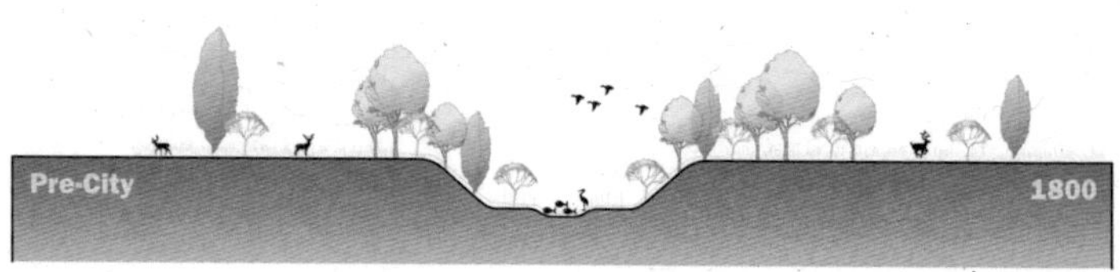

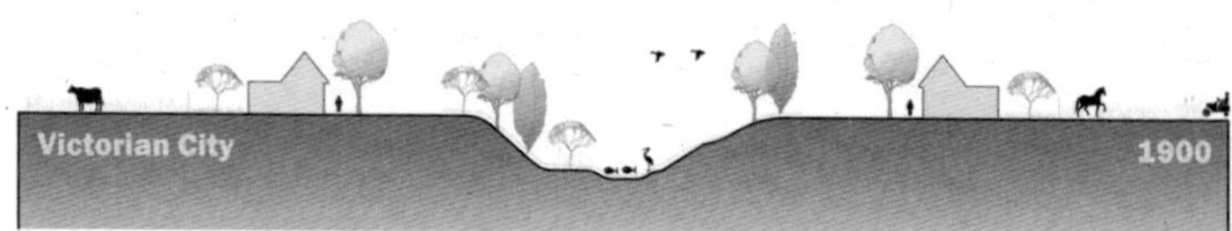

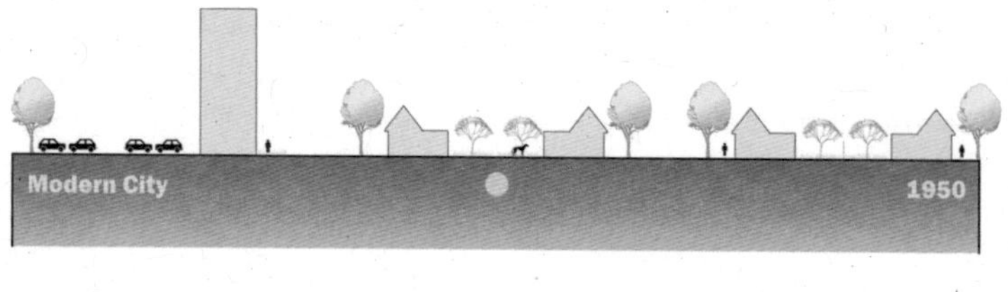

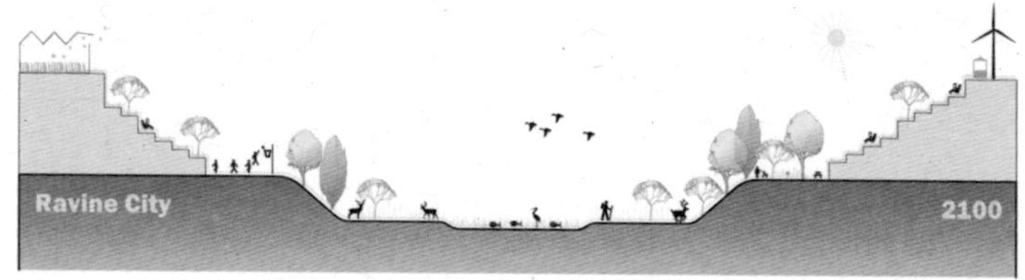

Above Sectional view of Toronto's ravines through history, with a final view diagramming the Ravine City concept.

Opposite 1. Toronto's original river system. 2. The unchanneled rivers of Toronto remaining on the surface. 3. Toronto with fully restored and enhanced ravines.

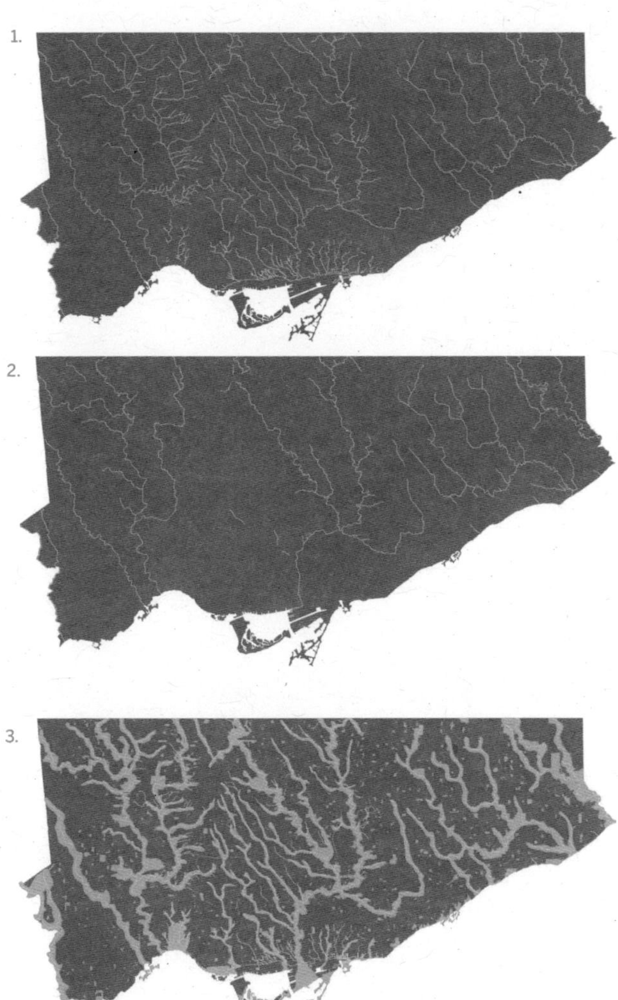
1.
2.
3.

Conventional roofing systems are expensive to maintain; roofs are exposed to the sun and undergo dramatic variations in temperature daily. Next to roads, roofs form the largest impervious rain-blocking surface in our cities. Ravine City addresses both of these problems by making its roof system hydrologically active and connecting it to ecological networks and public spaces. This new kind of roof becomes a productive, rather than negative, part of the urban fabric.

If we were building a new city from scratch it would be easy to integrate nature and green technologies, but we do not have that option. Retrofitting an existing city, like Toronto, is a bigger challenge. Raising homes, building infrastructure, and moving massive amounts of earth might not sound like an environmentally responsible strategy. But natural systems do not operate at the scale of a building or a park. They extend as a network to the scale of the city, the region, and the biosphere. So while continuing to support energy-obsolete, low-den-

Sectional view of the terraced housing showing green roofs and earth-sheltered housing.

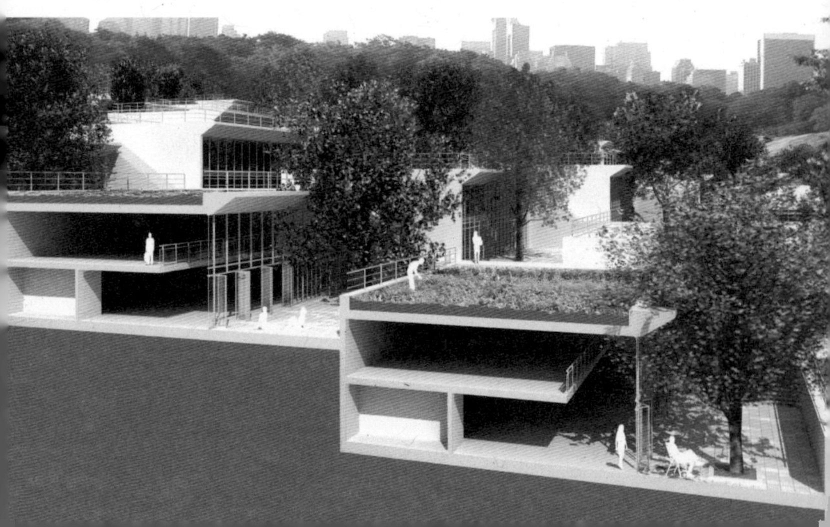

sity, single-detached-house neighborhoods is an unsustainable policy, constructing individual green buildings will not by itself produce a significant change in our environment. Projects on the scale of Ravine City are an investment in the future of our urban ecosystems: a self-sustaining green infrastructure.

In ecosystems, diversity is closely connected with network structure. A diverse ecosystem is resilient because it contains many species with overlapping ecological functions that can partially replace one another. The housing in the plan for Ravine City follows this model. Some houses are predominantly solar generators; others are wind generators, urban farms, waste water treatment devices, storm water retainers, urban forests, or recreational areas. Each unit generates a surplus of nutrients, energy, or restoration and shares its surplus with the community; these interconnections create a flexible and resilient urban network.

View across the garden terraces of Ravine City.

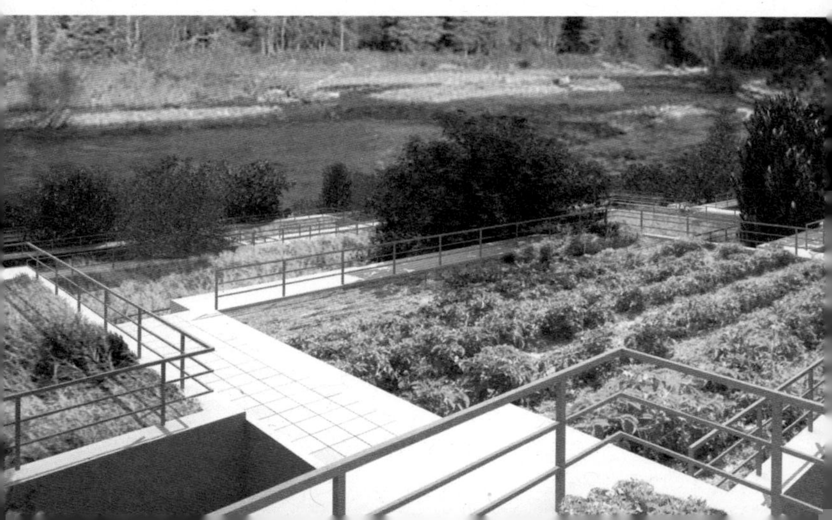

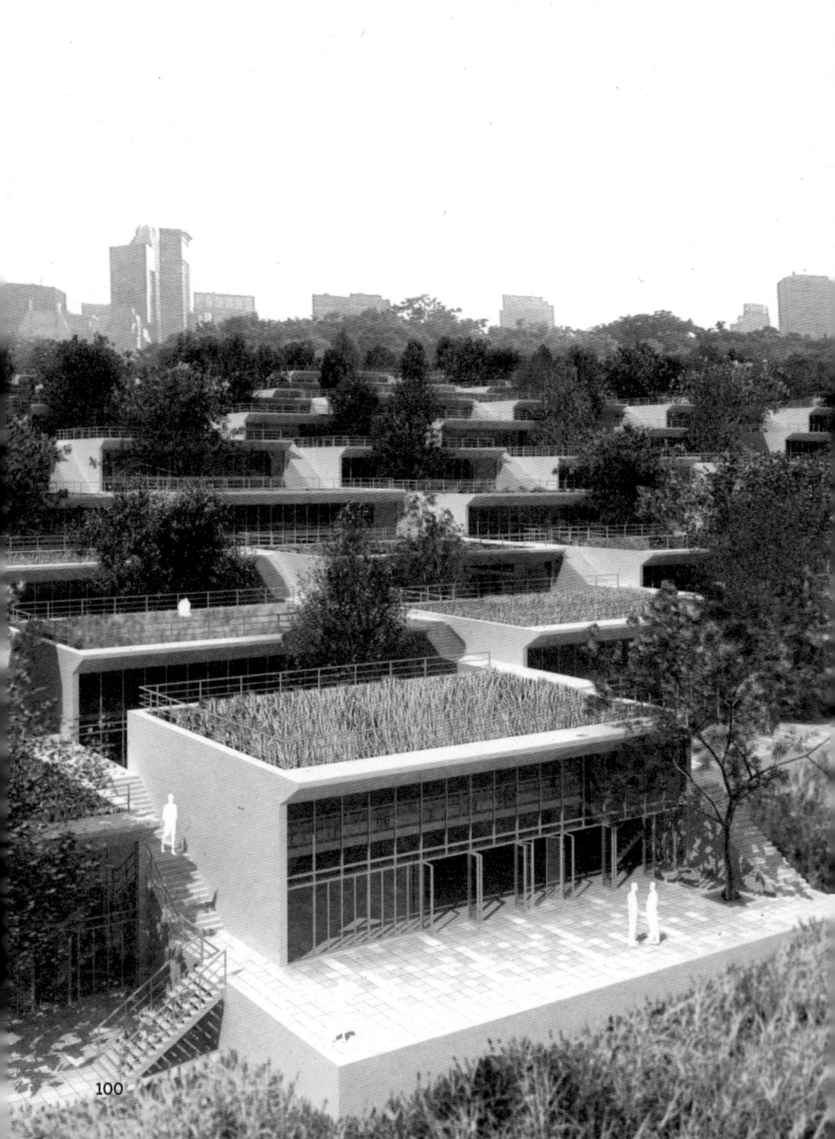

Toronto, like most modern cities, relies on linear dependent systems to support its needs. Linear dependent systems funnel resources without concern about their origin or the destination of wastes. Our urban sewage systems, for example, separate people from their wastes. Sewage is usually discharged downstream and its inherent fertility to farmland is lost.

This linear metabolic system is unnatural and unsustainable. Natural systems are cyclical: every output from a natural organism is also an input that renews and sustains the larger living environment. As our world becomes predominantly urban we need to adapt the design and planning of our cities to an ecosystemic model.

The word "ecosystem" derives its meaning from the Greek word *oikos* meaning "house, dwelling place, habitation." Ravine City connects the original meaning of oikos with ecosystems to propose housing that gives back to the system by participating in the flows of the larger network. In these projects urban outputs are also crucial inputs into urban production systems that treat and recycle productive wastes to generate energy and food.

Cities have become the forum where the future of our civilization and the biosphere will be determined. Our planet will not be able to accommodate ever-enlarging cities that continue to draw upon distant resources for their material needs and disposal of wastes. The alternatives described in the Ravine City project challenge the traditional urban image of the city as a collection of individual buildings and properties; in its vision of a new city, architecture and infrastructure are integrated with the landscape, and housing is connected to living social networks that follow the flow of natural systems. Ravine City makes natural cycles visible and brings nature into daily consciousness —it reconnects Toronto to the pleasure of its greatest natural resource —its ravines.

View of interconnected terraced housing alongside the ravines.

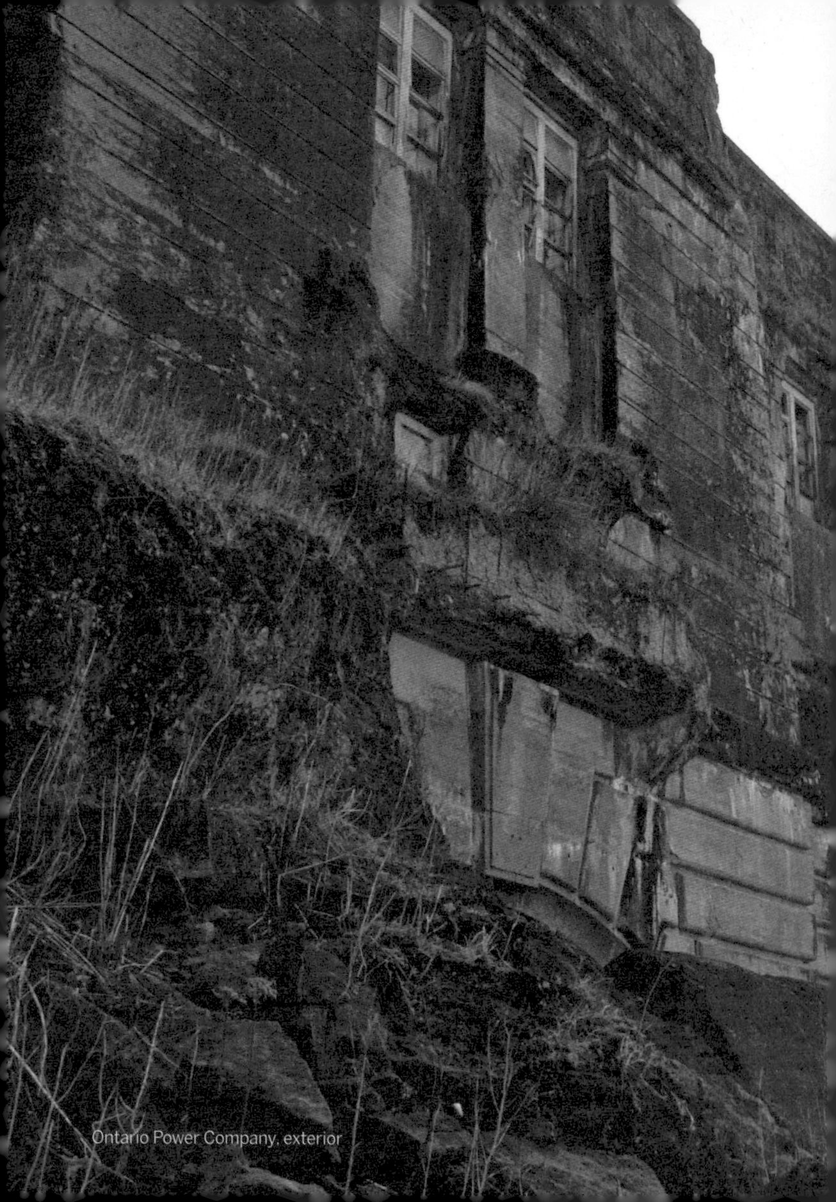

Ontario Power Company, exterior

In the Tunnels of the Ontario Power Company: Exploring the hydroelectric landscape of Niagara Falls

Michael Cook

In the summer of 2007, excavations across the road from Table Rock at the Horseshoe Falls briefly revealed for all to see one of the great industrial secrets of Niagara Falls.

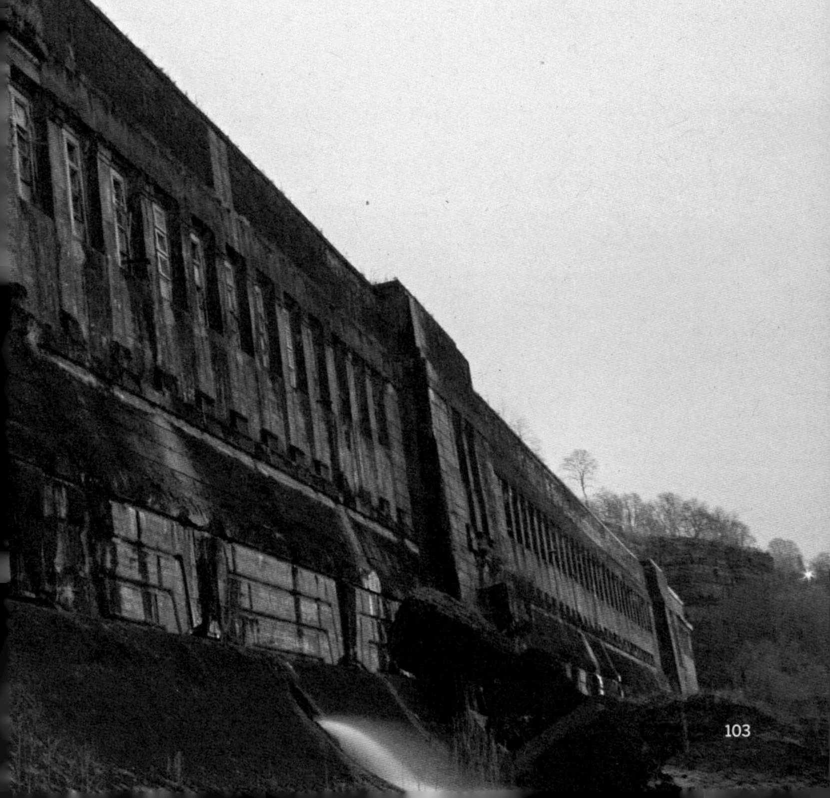

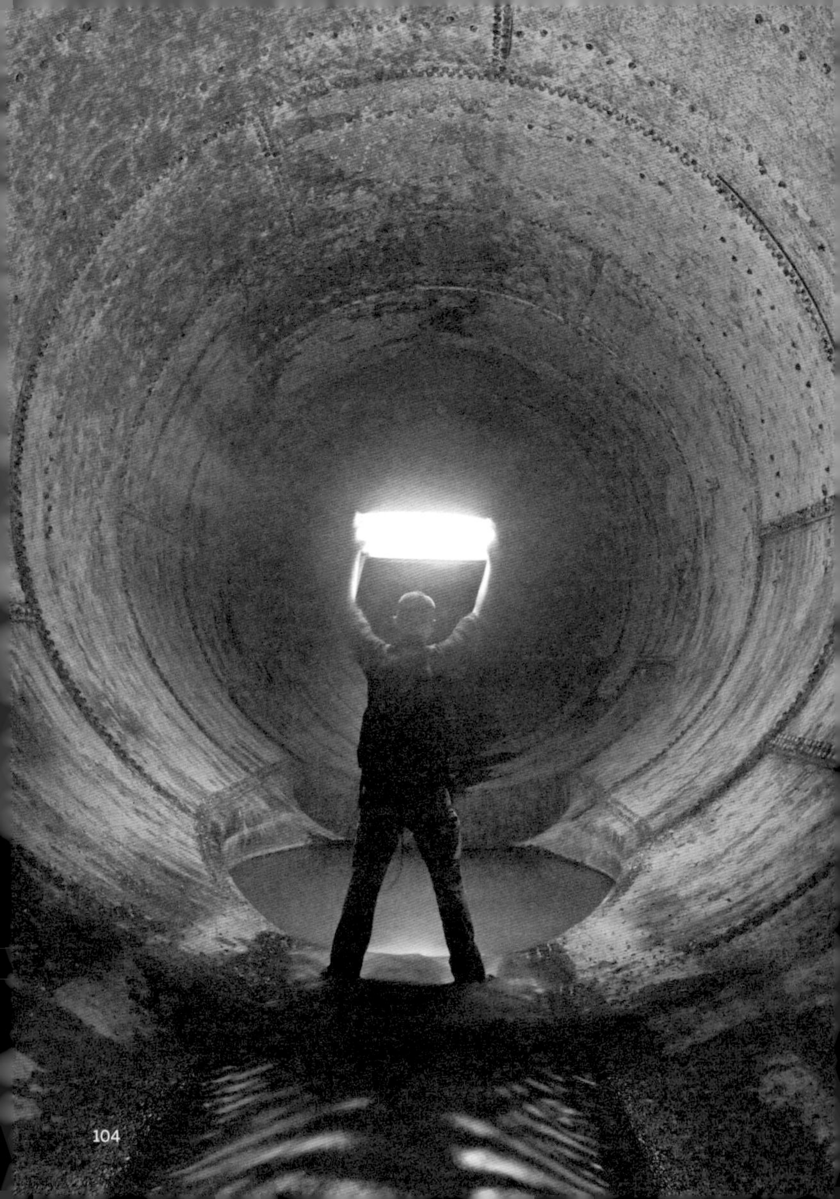

Beneath the lawns and parking lots that mediate the tourist's experience on the Canadian side, the parks are built upon great conduits of concrete and steel that once drew millions of liters of water to light and power Ontario and Western New York.

These tunnels lead not to either of the other early-century powerhouses that, shuttered, still lend their classic facades to the park experience, nor do they extend as the modern intake tunnels do to Queenston Heights and the Sir Adam Beck Generating Station. Instead, beneath surge structures whose rooftop lamps illuminate the falls, the conduits reach their own, underground cataract, a 54- meter drop through vertical penstock pipes to a powerhouse that sits empty and largely unnoticed at the bottom of the gorge. Built into the wall of the cliff, the building recalls in shape and spirit the mortuary temple of Hatshepsut at Luxor, but here the great river is at the building's feet, not three kilometers distant across desert and irrigated fields.

There are three tunnels, known as "distributors" to the people who once manned and operated the Ontario Power Company facility. The #1 Distributor is a riveted steel pipe 5.5 meters in diameter that, enveloped in concrete, drove the first turbines to come online when the plant opened in 1905. The # 2 Distributor was laid alongside it shortly thereafter, but is built entirely of concrete except for the steel walls at its terminus around the penstock openings. The third, built under conditions of wartime scarcity, was much smaller and clad in wood—it is not depicted in the photographs that accompany this article.

There is a sense of enormous size and power here, in tunnels that by the architectural standards of the period are fairly mundane. Two sources for this sensation can be easily expressed: immersion and verticality. For nearly a century, these spaces were full of water moving at 5 meters per second, and the sense of immersion they still projected was startling when I explored their lengths in 2006, about ten years after they had been dewatered. At that time, a small amount of river water and sediment still seeped through the original control gates into the conduits, and along with infiltrating groundwater this was

1 Distributor, penstock openings

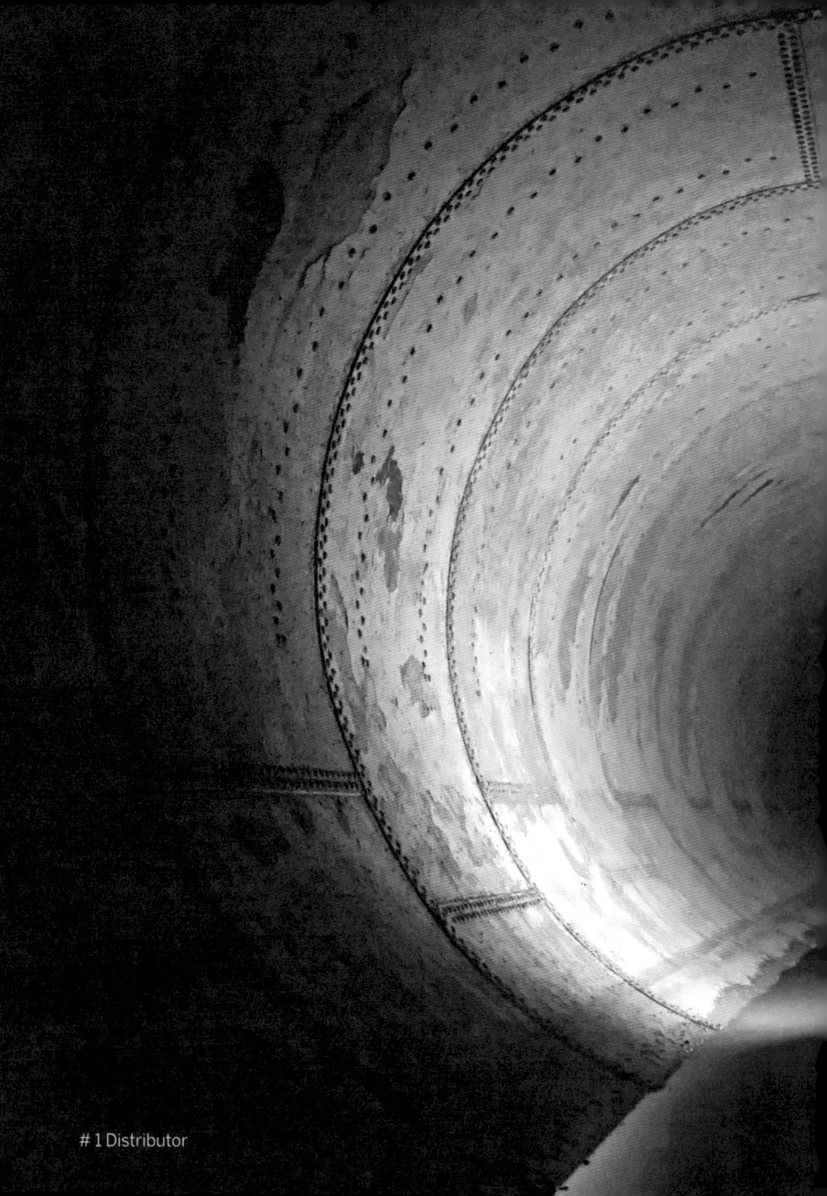

1 Distributor

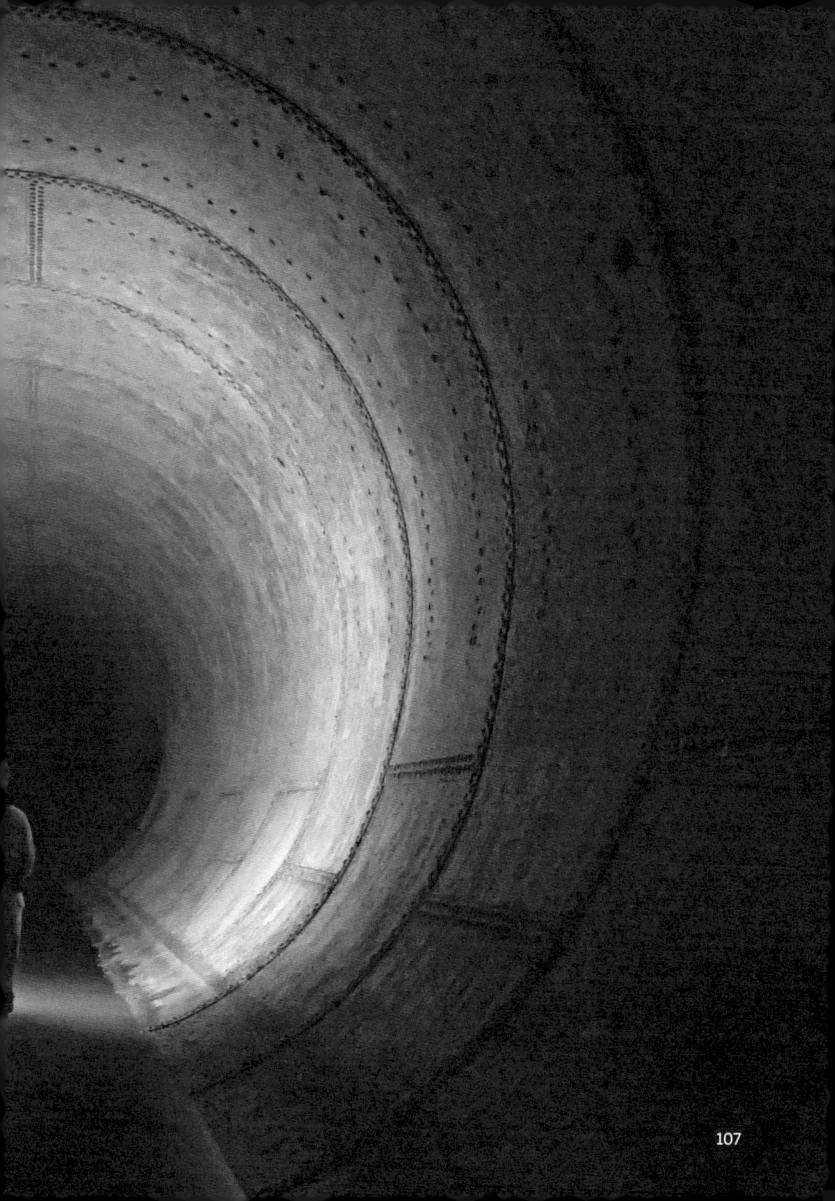

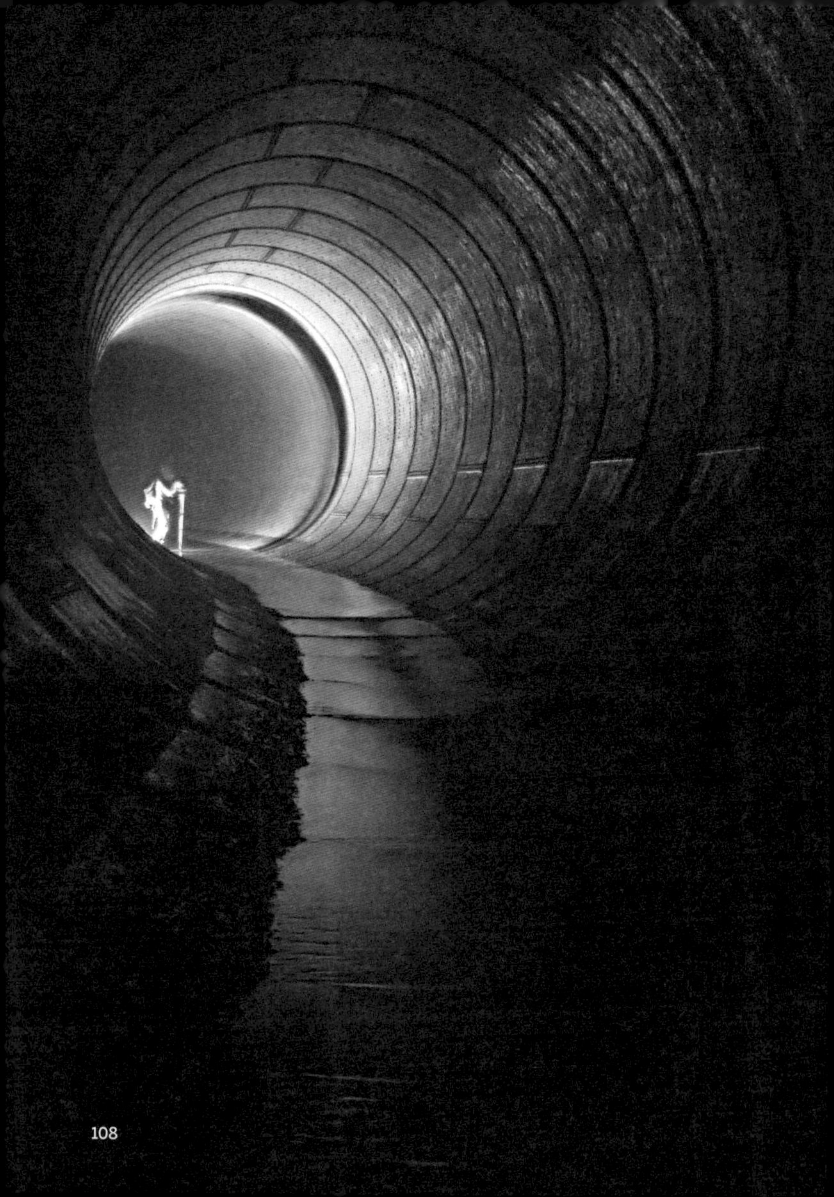

just enough to sustain a marginal ecosystem that included thousands of mussels and a few mudminnows. Though the space was now mostly empty, the sensation of far greater volumes of water remained in the air.

Reaching the penstocks, the vertical nature of the enterprise asserts itself. Dropping from the distributor, each penstock is three meters in diameter, and their empty caverns wait with a terrible thirst for the flow of the river. The first portal was full of water, its control valve sealed, but this pool then drained into an open second penstock, and beyond that four more iron mouths lay gaping in the subterranean murk. They wait for the undermining of structures built to constrain them, for the failure of the gatehouse's foundation or of its concrete bulkheads, the latter newly installed to replace the leaking gates and thus truly isolate the abandoned system from the river. While my colleagues and I have made our way into other hydroelectric tunnels at Niagara in the face of a certain amount of objective risk, none of us sought to chance fate skirting around these openings to reach the true end of either distributor.

When the plant was operational, surging water that passed beyond the penstocks was forced upwards into spillways that then spiralled down inside the cliff and beneath the powerhouse to the river. Pull up satellite images of the park and you can look down into the surge structures through their open roofs. One night we climbed from the boulder-strewn shoreline of the river up into one of these spillways, desperately seeking traction as we ascended more than 60 vertical meters up the steep corkscrew of a metal pipe that was crumbling to rusty grit. At the top, we stood beneath the stars while, unseen, the summer-night bustle of the park continued all around us.

This is the world beneath the park beside the falls. Much of it has been the subject of abatement in recent years: powerhouses gutted, forebays drained and filled in, tunnel entrances cemented over. Despite these efforts, parts of the hydroelectric heritage of Niagara Falls remain, a parallel world to the city's tourist landscape, buried and waiting for rediscovery.

2 Distributor, mudminnow, *Umbra limi*

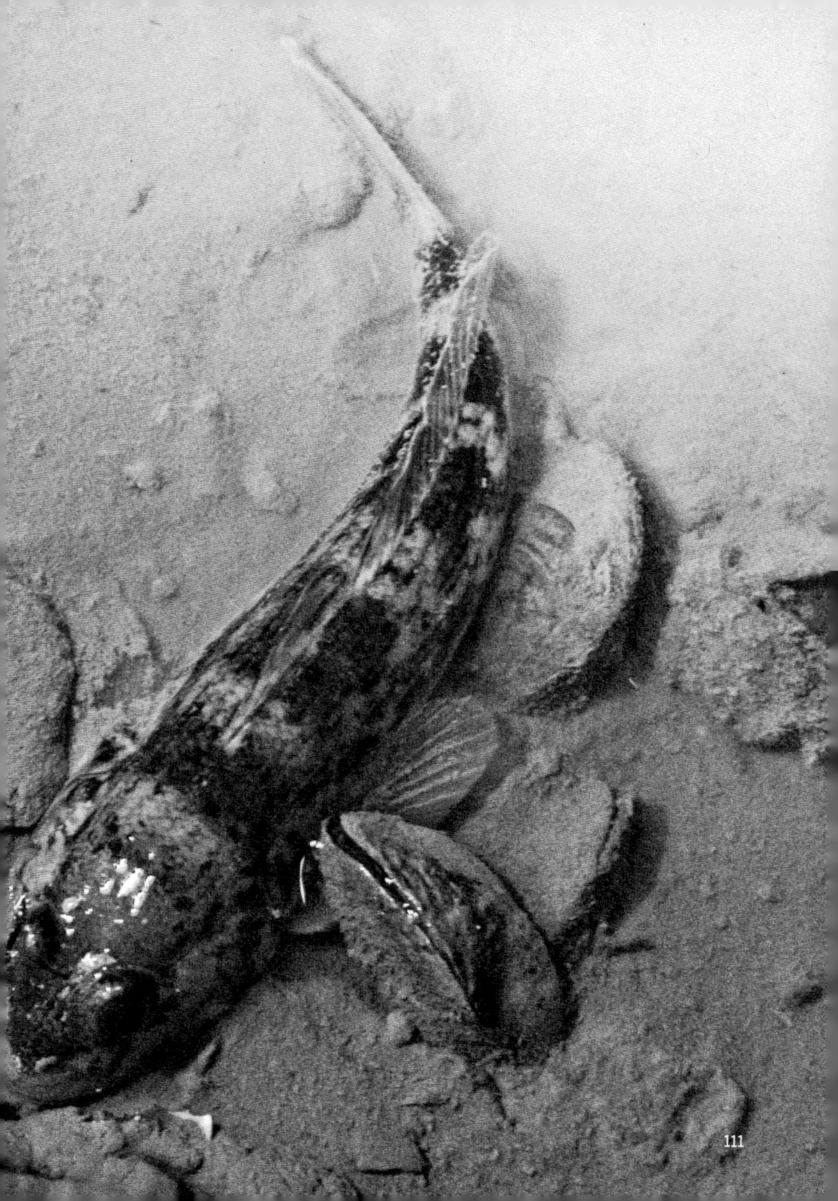

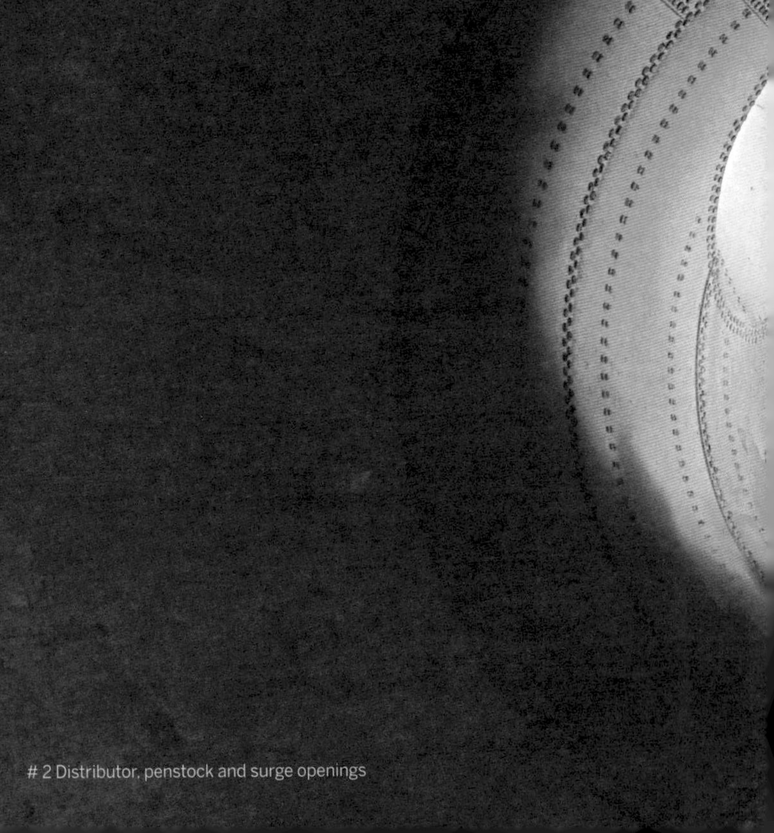

2 Distributor, penstock and surge openings

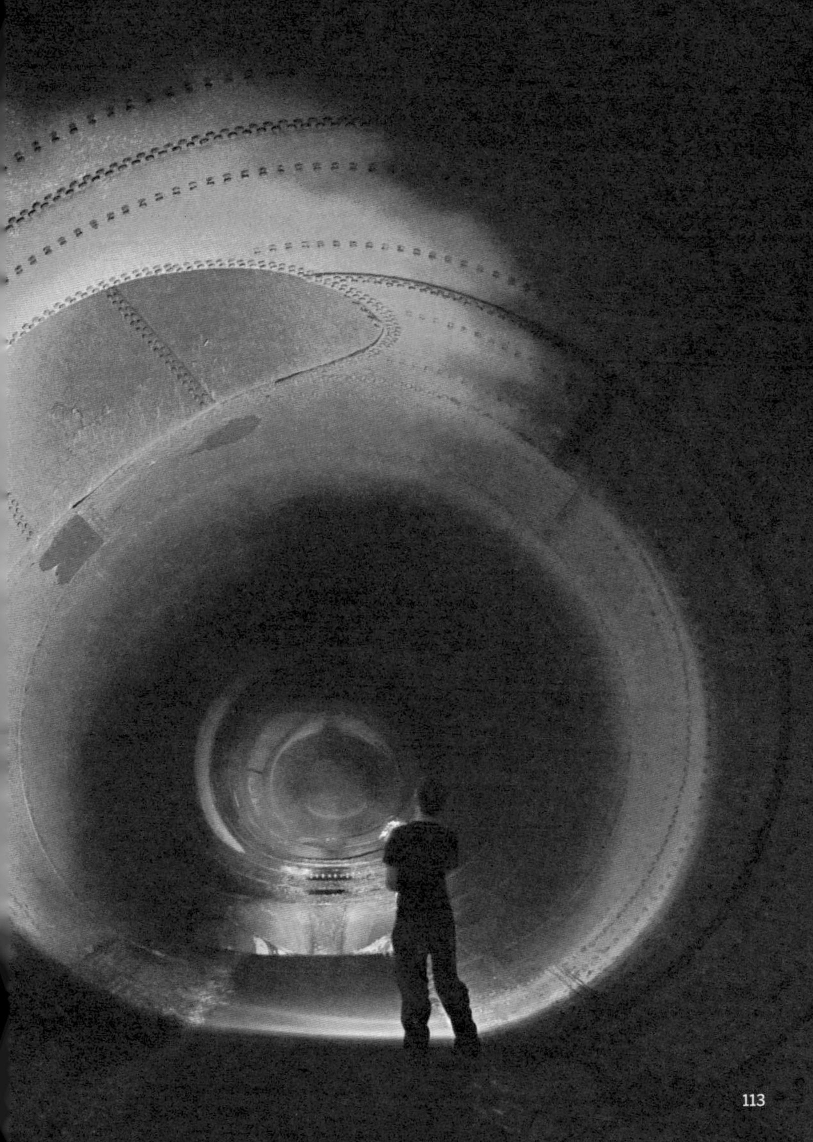

On Water and Development: A cautionary, microcosmic tale for a watershed near you

Robert Kirkbride

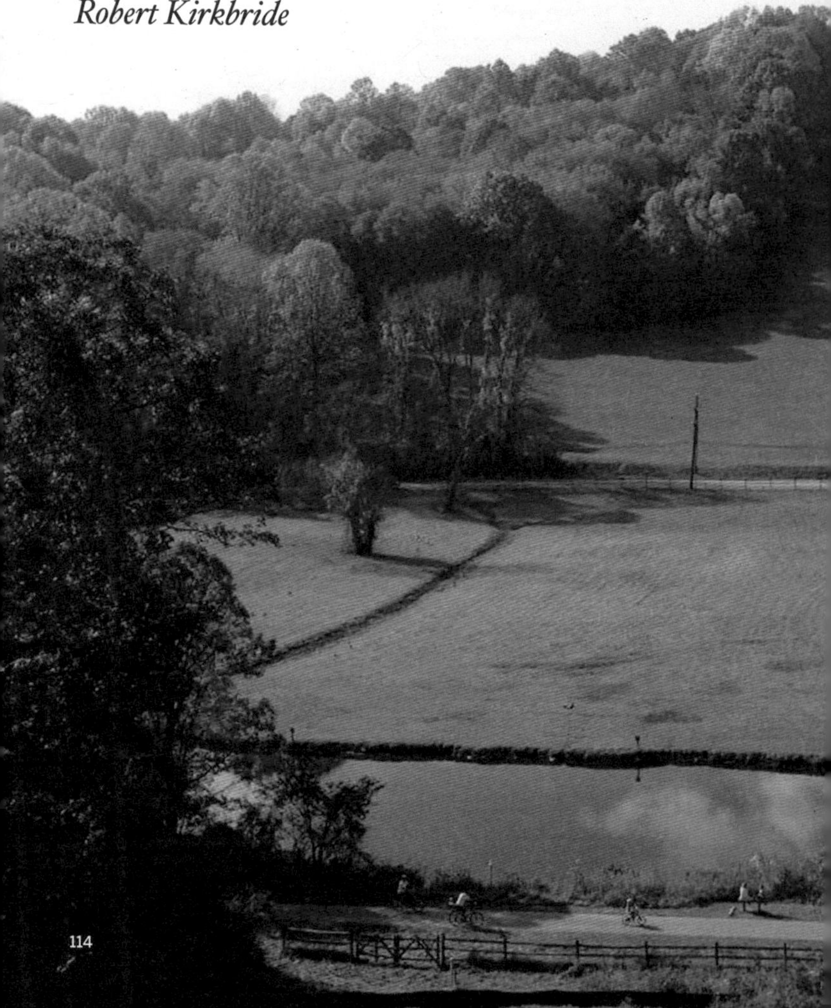

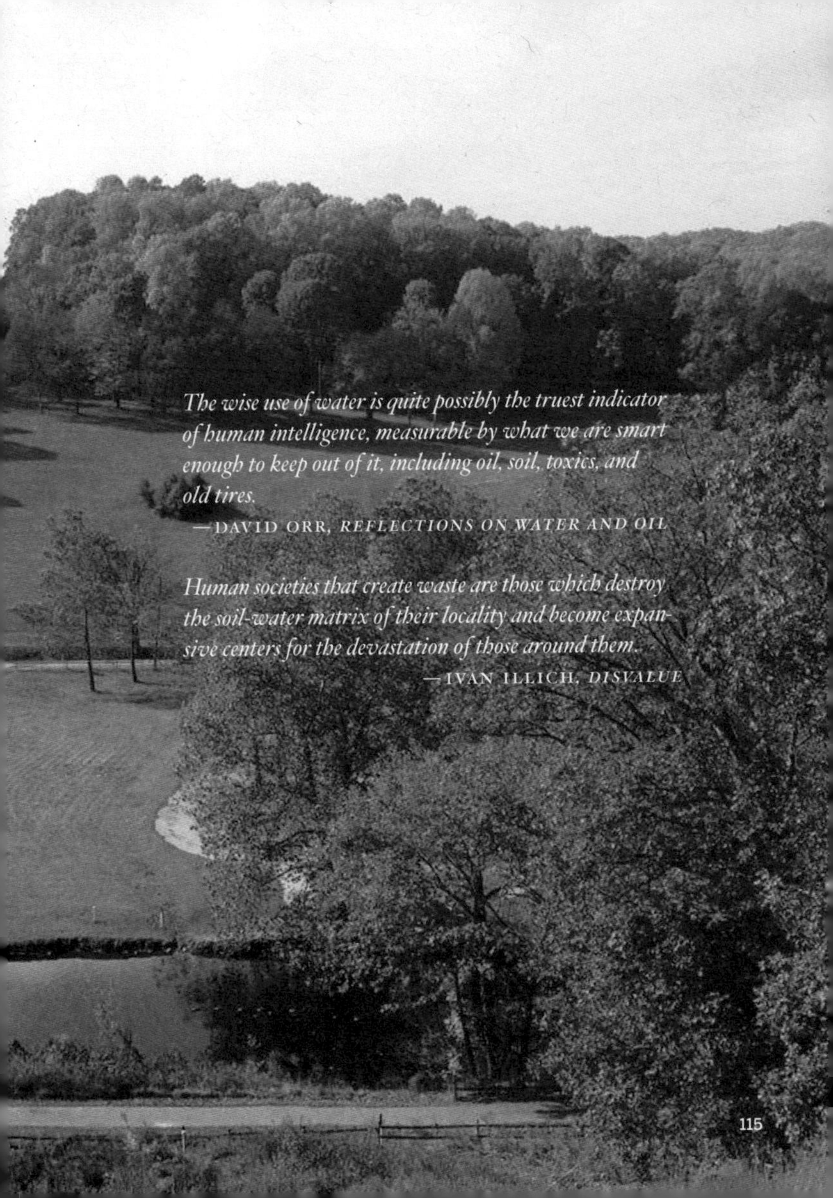

The wise use of water is quite possibly the truest indicator of human intelligence, measurable by what we are smart enough to keep out of it, including oil, soil, toxics, and old tires.

—DAVID ORR, *REFLECTIONS ON WATER AND OIL*

Human societies that create waste are those which destroy the soil-water matrix of their locality and become expansive centers for the devastation of those around them.

—IVAN ILLICH, *DISVALUE*

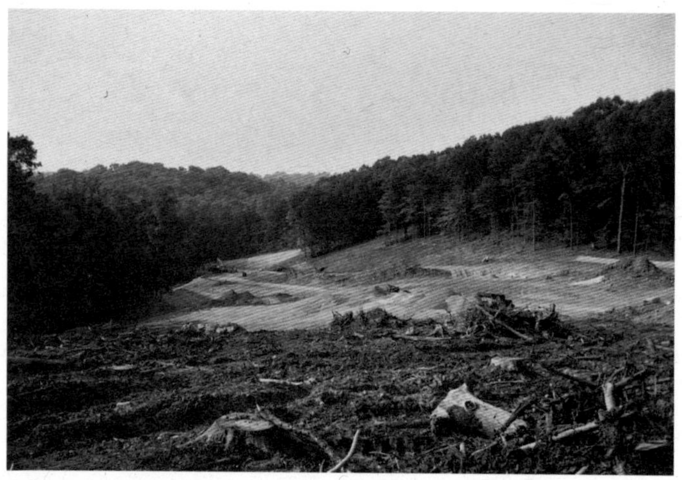

There are over 320,000 kilometers of underground pipeline for natural gas and hazardous liquids in the US, Canada, and Mexico, each intersecting countless large and small bodies of water, disrupting and threatening watersheds, nullifying landowners' rights, and casting severe doubt on the wisdom of our utility infrastructure and land practices. At one location, where 900 meters of this system intersects with a small, forested site in the mid-Atlantic region of the United States, human and environmental forces are combining to create a precarious yet increasingly common situation.

1. We begin underground, in the soil. According to the Department of Agriculture Survey, this site has well-drained soil, with Glenelg and Brandywine stone loam occupying ridgelines and moderately steep

Above Four hundred meters upstream, a 300-meter clearcut for site development intersected the transcontinental pipeline corridor. Although the approved plans indicated limited tree removal, this stipulation was not enforced by local authorities.

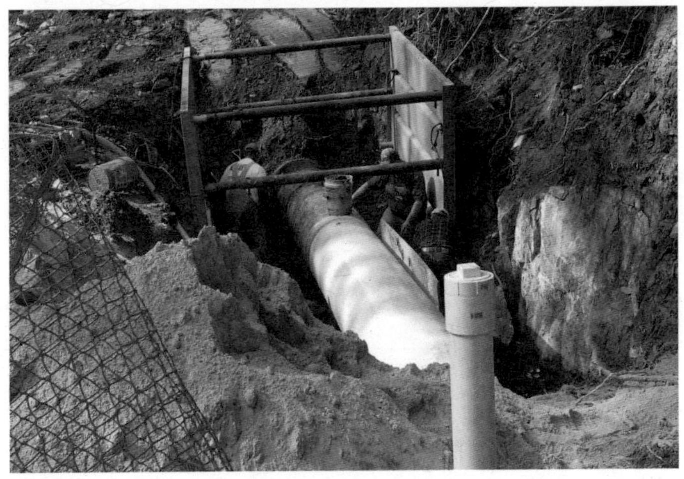

slopes, leading down to the Wehadkee silt loam that constitutes the banks and alluvial confluence of an adjacent stream, Ludwig's Run, with a broader creek, the east branch of the Brandywine. Composed of a thin mat of leached leaf mold, weathered granite, gneiss and mica schist, as well as the occasional infusion of decaying wildlife, the earth readily absorbs rainwater and is moderately fertile. Sporadic large boulders of sandstone, known as floaters, make cultivation unfeasible, and so periodically the site's timber is selectively harvested. Overhead, the black, red, white, and chestnut oak forest is rapidly shifting to beech, ash, and hickory with an understory of red maple, dogwood, American chestnut, elm, and sassafras. This lower, messier zone of the forest is a valuable habitat for resident and migratory birds, including cardinals, bluebirds, woodthrushes and hummingbirds. There are also red fox, box turtles, mountain laurel, seventeen-year cicadas, and cicada wasps. We are in the East Brandywine watershed, which empties ponderously southeastward into the Delaware River.

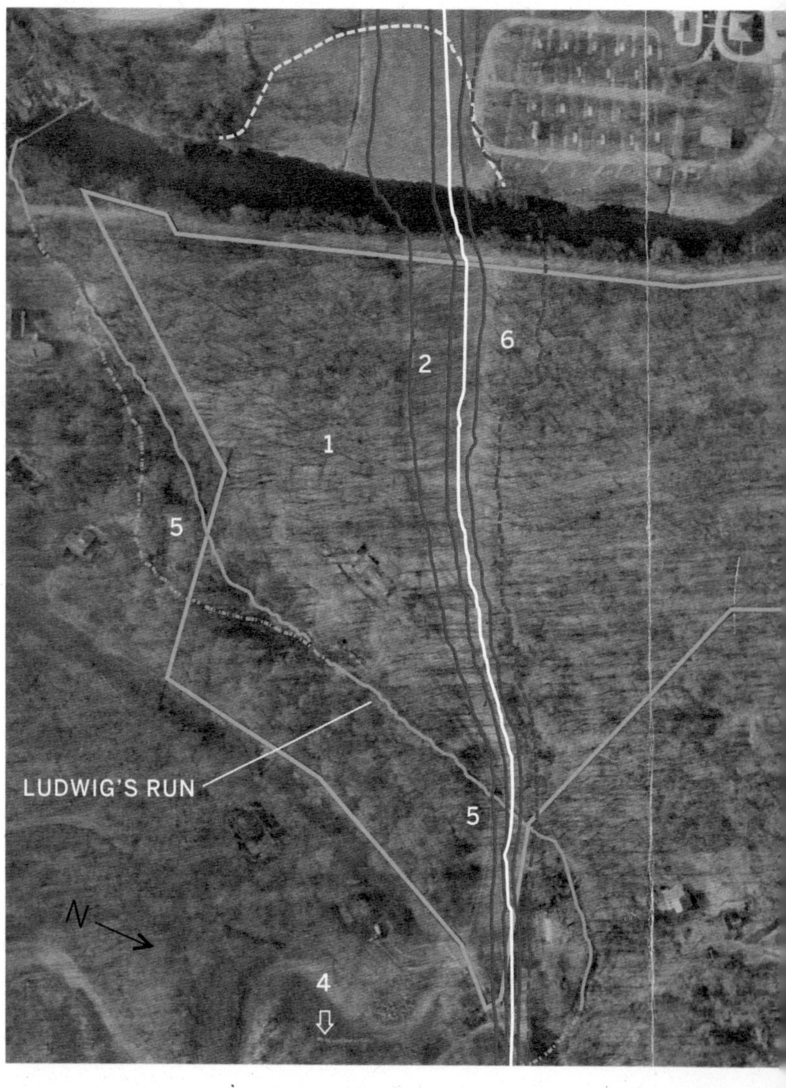

6
2
1
5
LUDWIG'S RUN
5
N
4

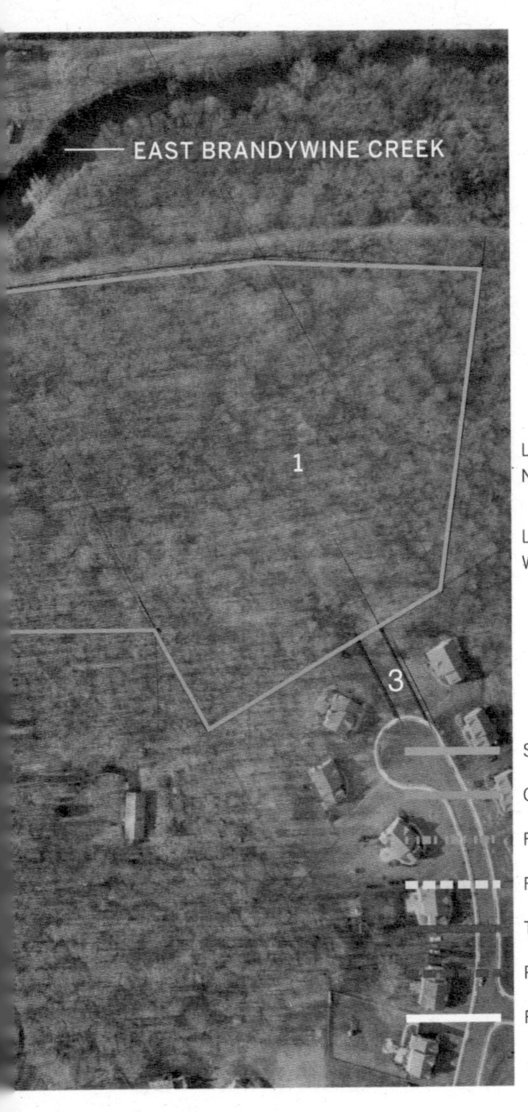

LATITUDE
N40.0278545

LONGITUDE
W75.69719091

SITE BOUNDARY

CURRENT STREAMBED

FORMER STREAMBED

FLOOD LINE

TRANSCONTINENTAL PIPELINES

PIPELINE EXPANSION

FIBEROPTIC LINE

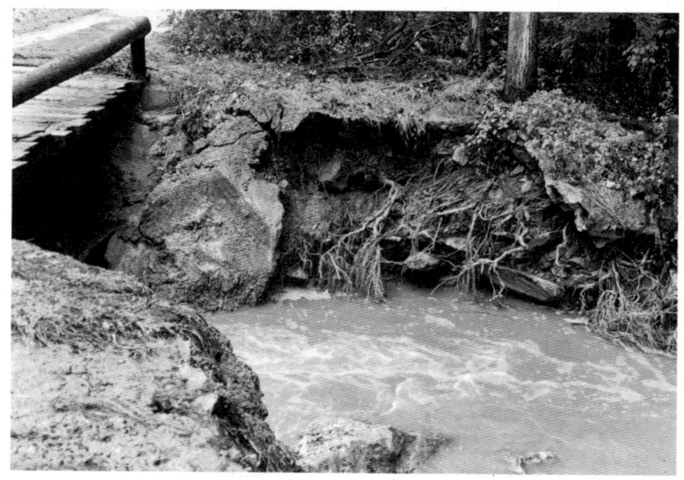

Over the past sixty years, the flow of water through this property has been increasingly disrupted by converging global, regional, and site-specific forces.

2. Since 1950, three transcontinental natural gas pipelines and one fiberoptic bundle have been embedded in the eastern portion of the site in an easement ranging from 23 to 32 meters wide. Averaging 86 cm in diameter and just under a meter below grade, the pipelines are part of a continuous utility corridor connecting New York City with the Gulf of Mexico. Above ground, much of this easement has been maintained by the pipeline company as a meadow that, like other railway and highway corridors, has become a conduit for the rapid dispersion of invasive vegetation. At water's edge, the picture is more disturbing.

Above Downstream, erosion to streambanks and the threat of damage to private property necessitated state-funded site remediation.

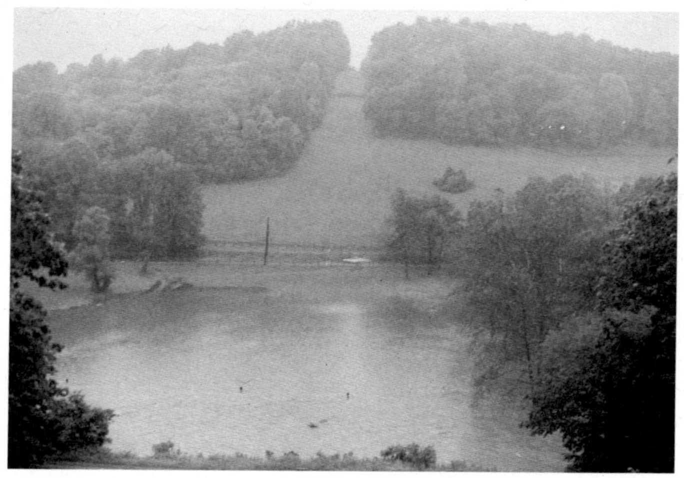

3. Fifty years of upstream development has taken a toll on site hydrology. Ludwig's Run, formerly a seasonal brook, is now a year-round stream and has repeatedly "jumped" from its bed due to intensifying storms and faulty stormwater management practices higher in the watershed. The earliest-built communities upstream included few or no stormwater controls. More recently, developers have followed local ordinance, yet their predisposition is to costly, over-engineered solutions that manhandle water by reshaping rather than complementing local site conditions, and their measures have not protected the stream corridor from the increased hydraulic flows they introduced.

4. Between 2003 and 2005, human hubris and extreme weather combined to threaten private and public infrastructure and potable water. Whether the series of traumatic "five-hundred-year" or "thousand-year" storms reflected broader global patterns or "acts of God"—a phrase heard frequently from engineers and officials—these events were magnified by unrestricted clearcutting and inadequate site man-

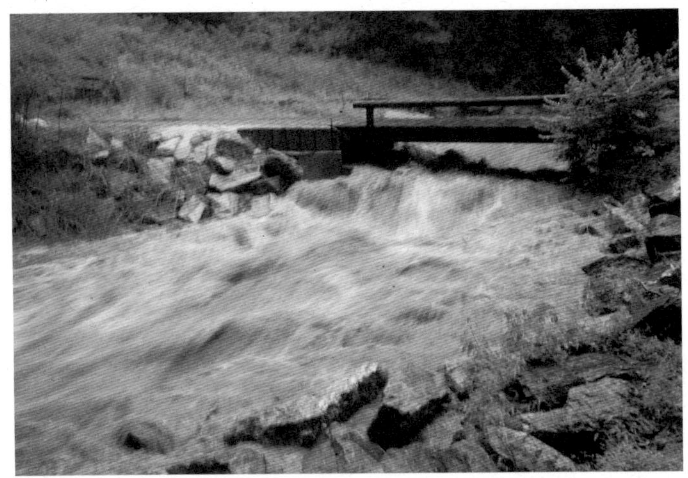

agement in new construction immediately upstream, causing severe erosion to banks, lowering the streambed dramatically, and risking environmental devastation by exposing the natural gas pipelines to the possibility of rupture in the stream. When local residents presented documentation of the cumulative failures of the recently built hydraulic features, officials offered the boilerplate response that the controls were designed and implemented according to approved standards. Unfortunately, just as the map is not the territory, the "approved standards" do not necessarily apply for a given soil type or site topography.

5. During this period, the Department of Environmental Protection and County Conservation District proactively provided emergency permits to homeowners for (repeated) bridge repairs, and coordinated a state-funded stream remediation project. Fortunately, "Band-aid" solutions by Williams-Transco were rejected by the DEP in favor of lowering the pipelines and employing low-impact features, such as cross vanes.

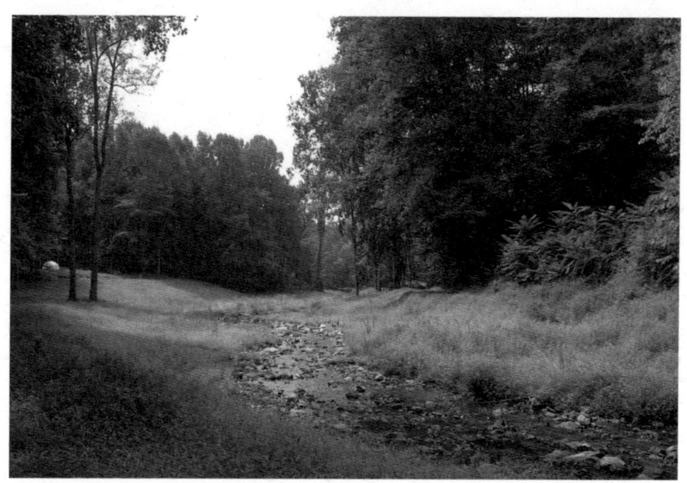

6. Nonetheless, despite the US$250,000-plus spent by state agencies and individual owners to restore balance to this degraded waterway, upon completion of the remediation Williams-Transco immediately began efforts to expand its easement to add a new line. Subsequent pushback from landowners and authorities has convinced pipeline officials that it would be expedient to replace an existing 76.2 cm natural gas pipeline with a new 106.6 cm line. However, current plans call for removal of more than 1,500 trees, renewed disturbance to this fragile segment of Ludwig's Run, and the temporary rerouting of the East Brandywine Creek less than 1.6 km above the drinking water intake for residents downstream. Although serious flaws have been found in the plan, and objections raised by government agencies, owners, and elected local and state officials, the pipeline company has declared its intent to employ eminent domain to achieve its objective, increasing private gain under the guise of "public use" and at the expense of the public good.

Research assistants: Hironao Kato and Richard Yeh

Washed Water *Angela Grauerholz*

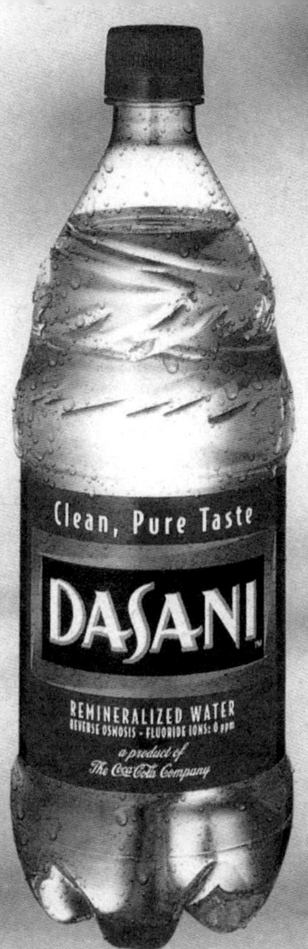

INTRODUCING WASHED WATER

ou may not realize it, but not all water is created equal. New DASANI uses a state of the art everse Osmosis process for purity which removes organic compounds, chlorine and bacteria. hen, just the right amount of minerals are added for a clean, pure taste. So you'll love what's in and what's not.

Life Simplified

NI", "Washed Water" and "Life Simplified" are trademarks of The Coca-Cola Company.

Pure and generic,

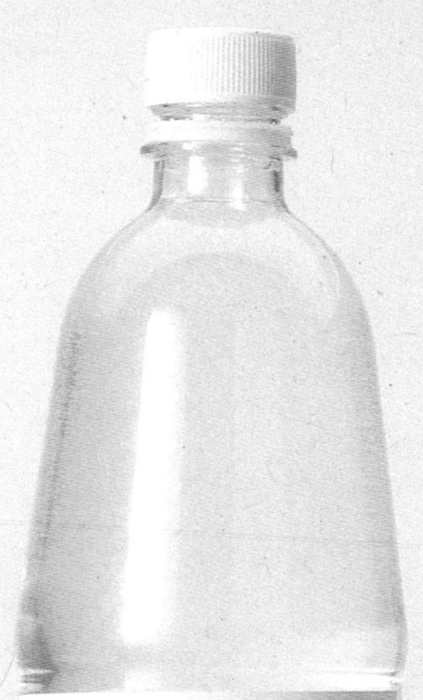

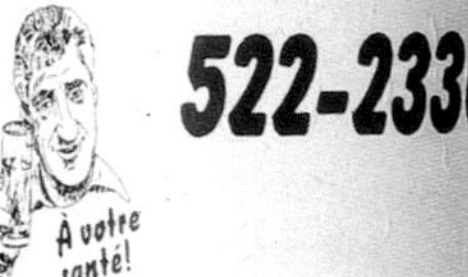
MONSIEUR EAU
EAU PURE, OBTENUE
PAR DISTILLATION
522-2336
À votre
santé!
RÉSIDENTIEL ❖ COMMERCIAL ❖ INDUSTRIEL

plentiful and available,

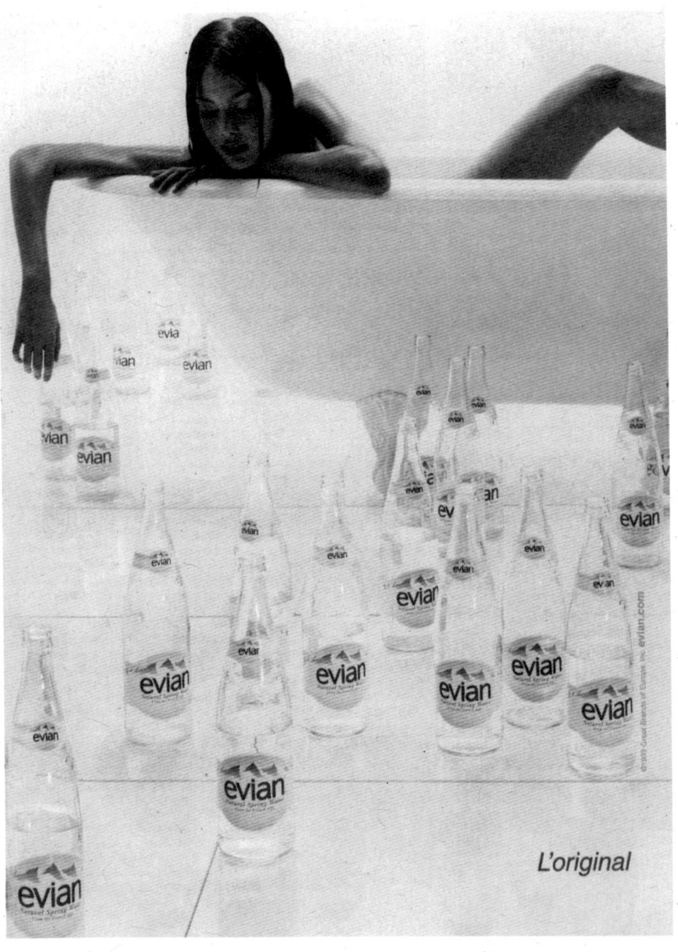

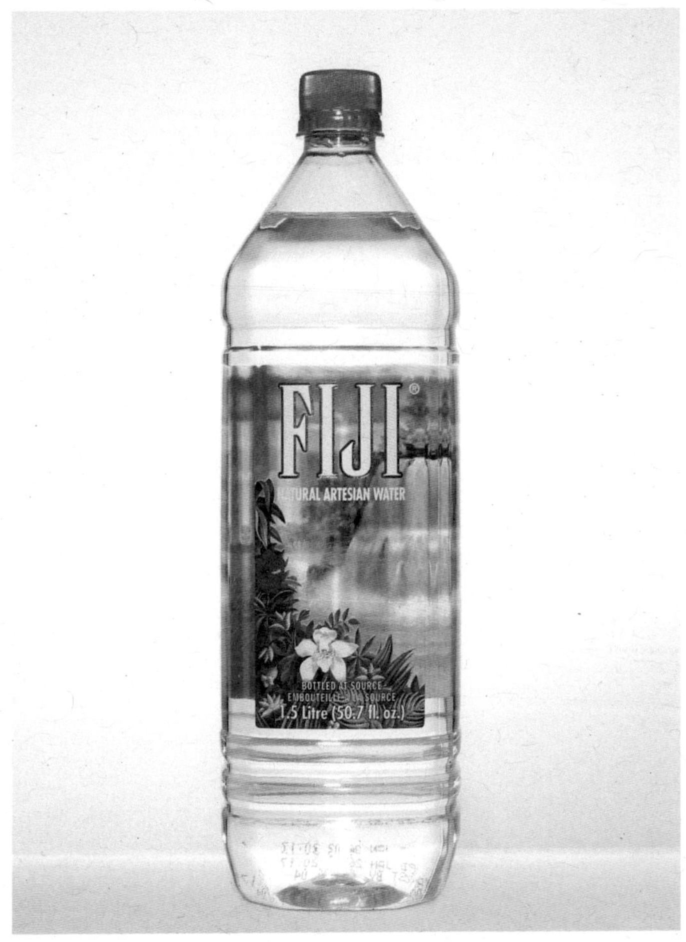
FIJI®
NATURAL ARTESIAN WATER
BOTTLED AT SOURCE
EMBOUTEILLÉ À LA SOURCE
1.5 Litre (50.7 fl. oz.)

portable and international,

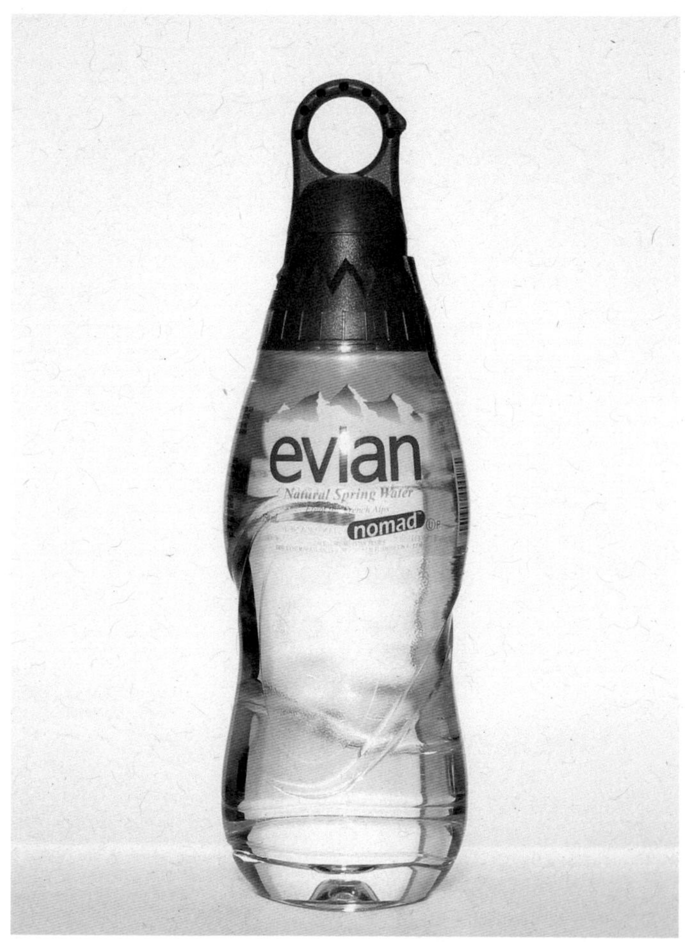

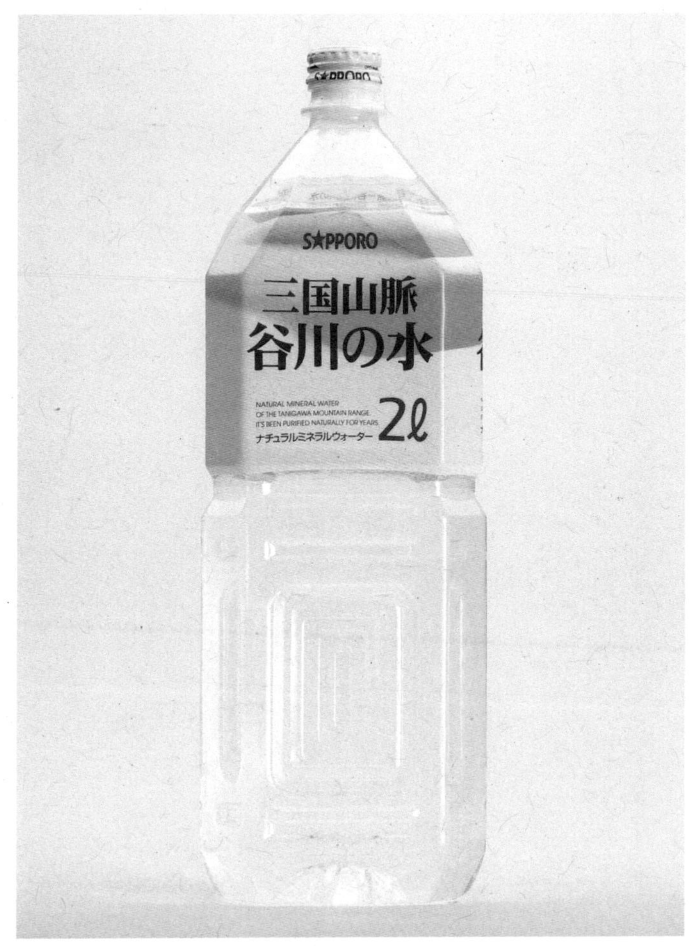
SAPPORO
三国山脈
谷川の水
NATURAL MINERAL WATER
OF THE TANIGAWA MOUNTAIN RANGE
IT'S BEEN PURIFIED NATURALLY FOR YEARS
ナチュラルミネラルウォーター
2ℓ

multilingual, transparent, and free.

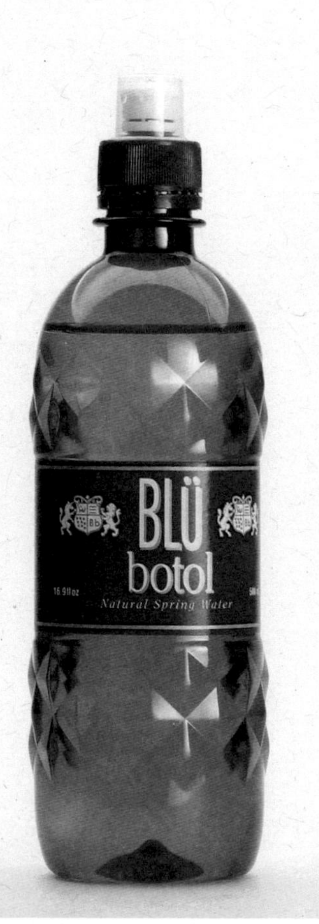

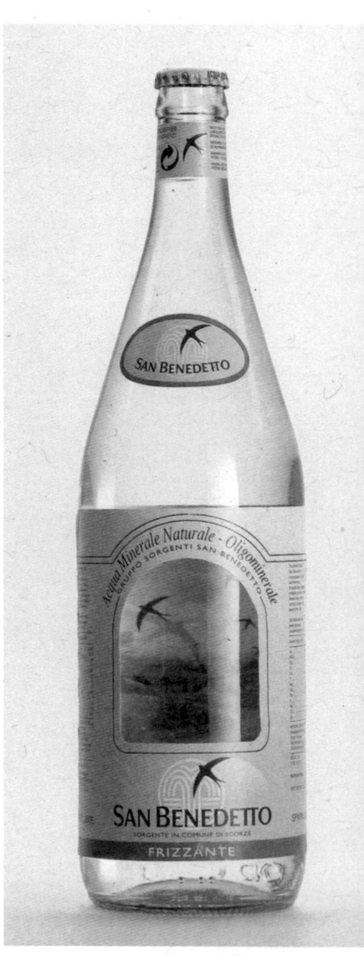

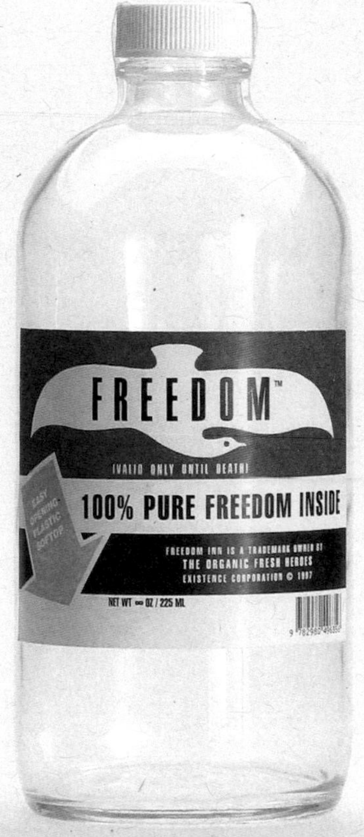
FREEDOM™
(VALID ONLY UNTIL DEATH)
EASY OPENING PLASTIC SOFTOP
100% PURE FREEDOM INSIDE
FREEDOM INN IS A TRADEMARK OWNED BY
THE ORGANIC FRESH HEROES
EXISTENCE CORPORATION © 1997
NET WT ∞ OZ / 225 ML

Traditional,

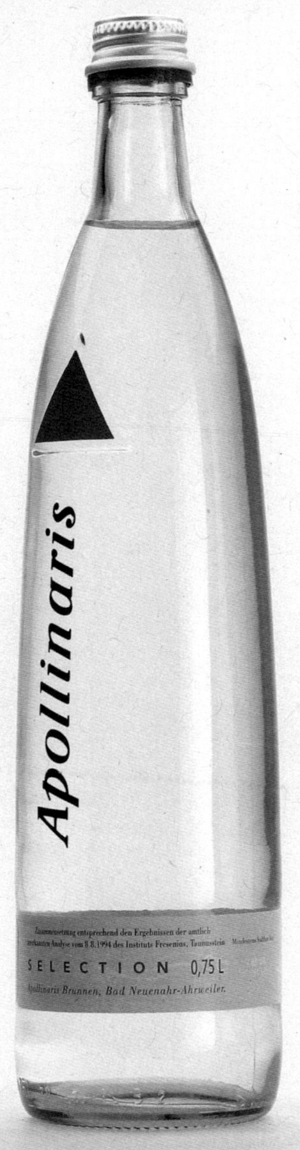
Apollinaris
entsprechend den Ergebnissen der amtlich
Analyse vom 8.8.1994 des Instituts Fresenius, Taunusstein
SELECTION 0,75 L
Apollinaris Brunnen, Bad Neuenahr-Ahrweiler.

healthful, enhanced,

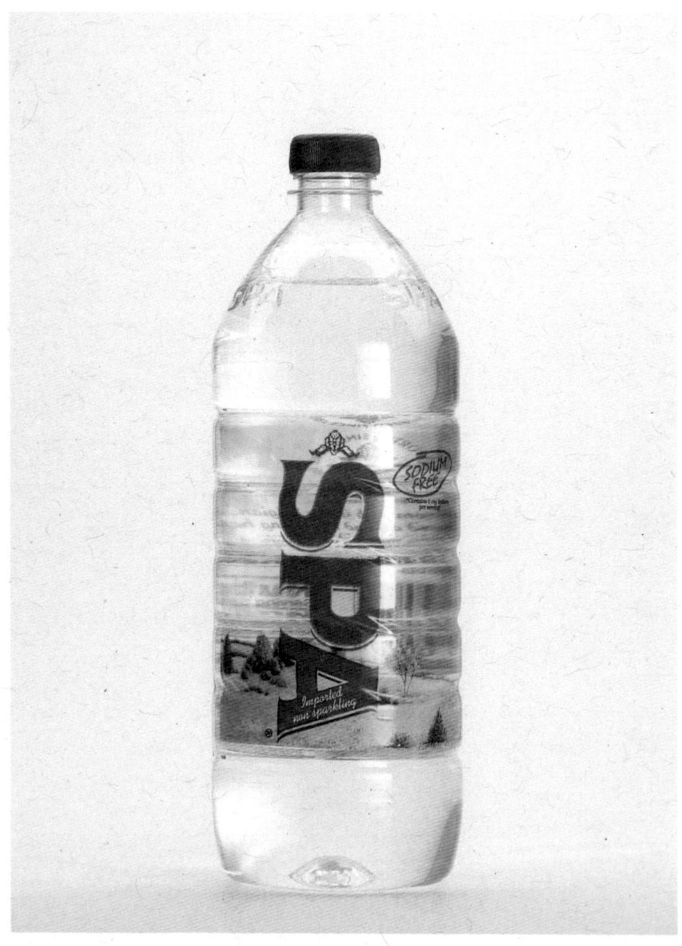

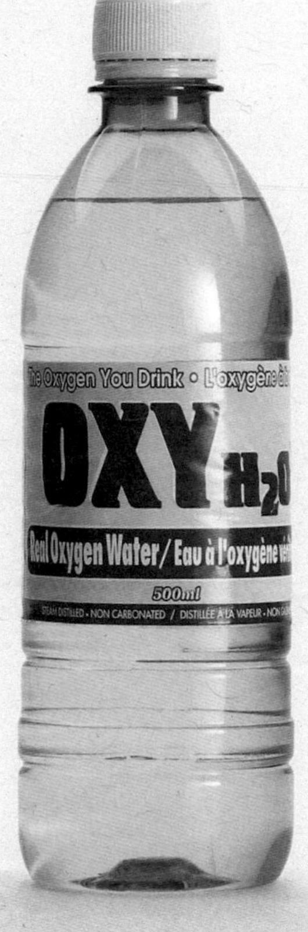
The Oxygen You Drink • L'oxygène
OXY H_2O
Real Oxygen Water/ Eau à l'oxygène
500ml
NON CARBONATED / DISTILLÉE À LA VAPEUR

sparkling or still,

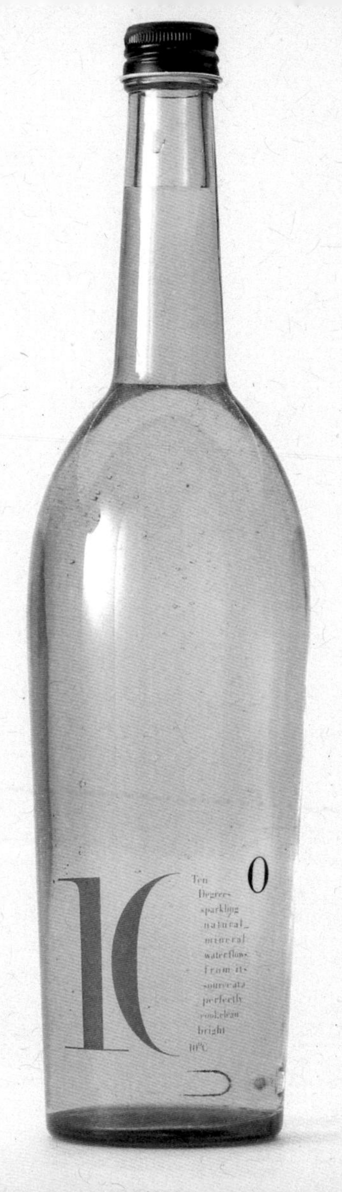
10
0
Ten
Degrees
sparkling
natural
mineral
waterflows
from its
source at a
perfectly
cool clean
bright
10°C

glacial and smart.

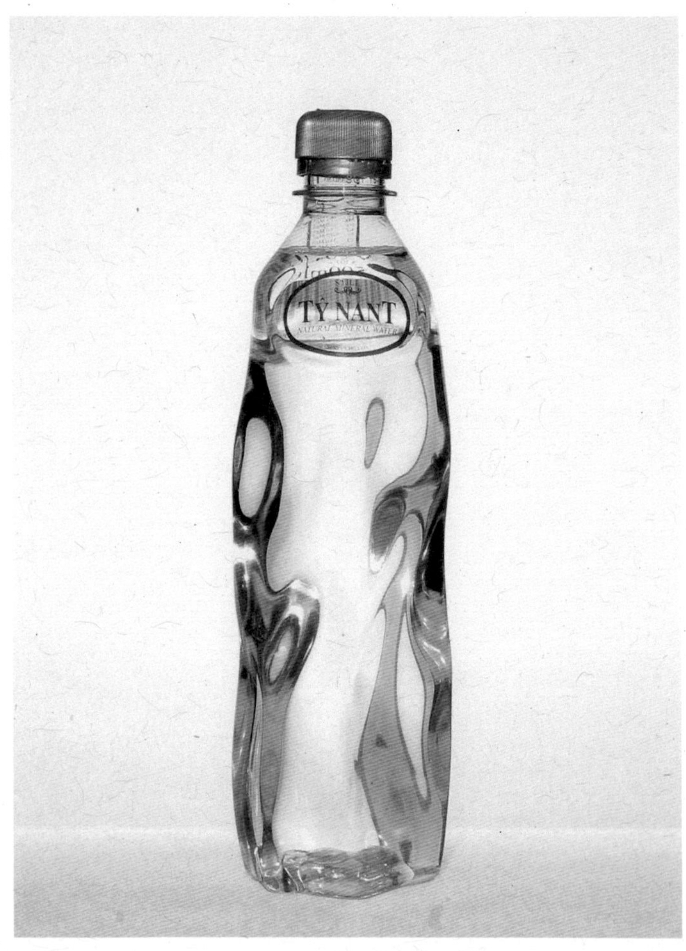

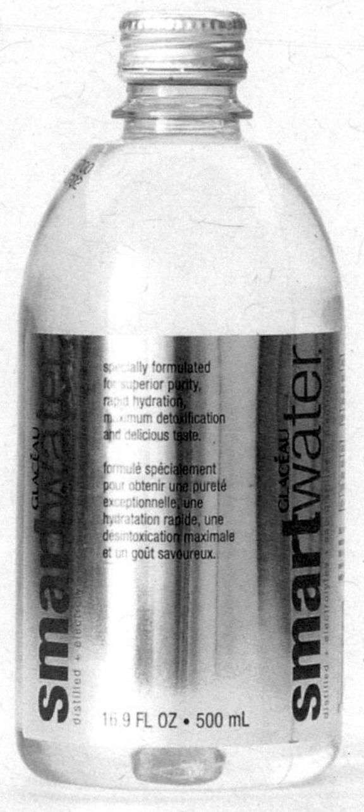
GLACÉAU
smartwater
distilled + electrolytes
specially formulated
for superior purity,
rapid hydration,
maximum detoxification
and delicious taste.
formulé spécialement
pour obtenir une pureté
exceptionnelle, une
hydratation rapide, une
désintoxication maximale
et un goût savoureux.
16.9 FL OZ • 500 mL

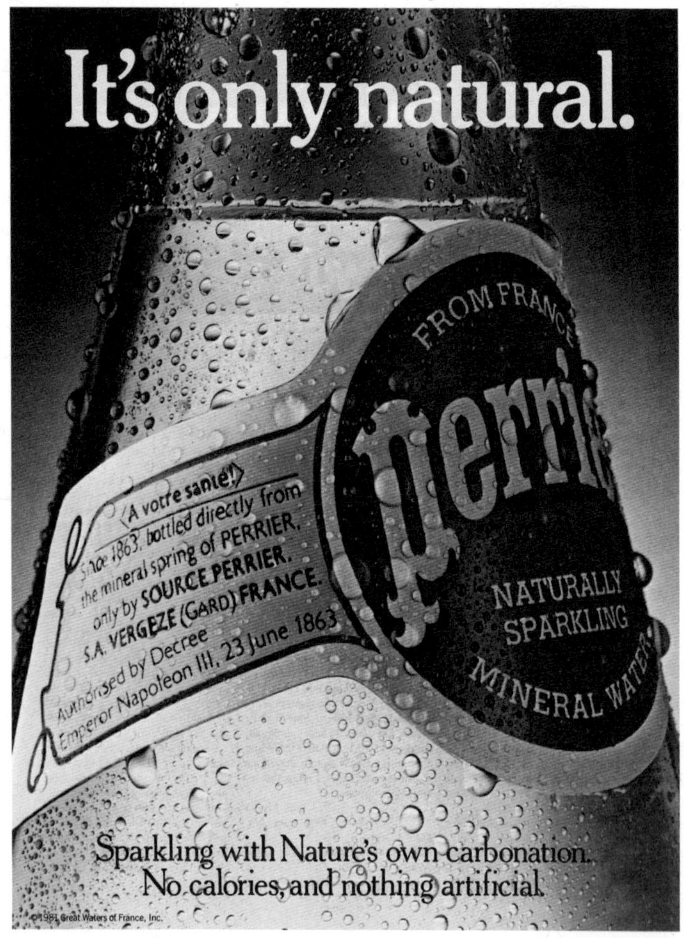
It's only natural.
FROM FRANCE
NATURALLY SPARKLING
MINERAL WATER
‹A votre santé!›
Since 1863, bottled directly from
the mineral spring of PERRIER,
only by SOURCE PERRIER,
S.A. VERGEZE (GARD) FRANCE.
Authorised by Decree
Emperor Napoleon III, 23 June 1863
Sparkling with Nature's own carbonation.
No calories, and nothing artificial.
©1981 Great Waters of France, Inc.

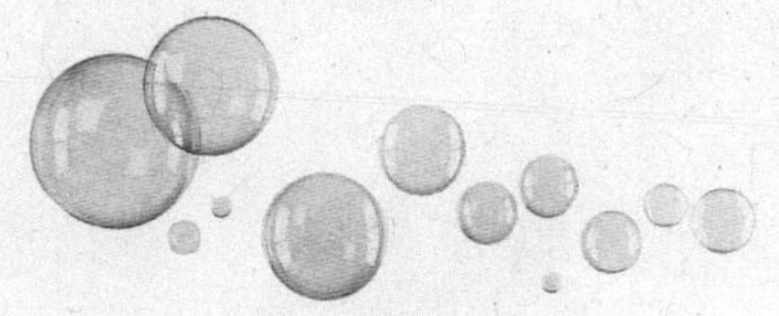

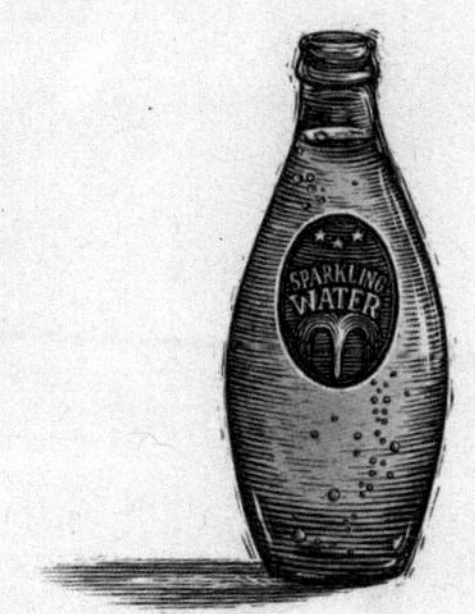

NOW THERE'S A TASTEFUL ALTERNATIVE. NO CAFFEINE. VIRTUALLY

"NutraSweet and the NutraSweet symbol are registered trademarks of The NutraSweet Company for its brand of sweetening ingredient and designed.

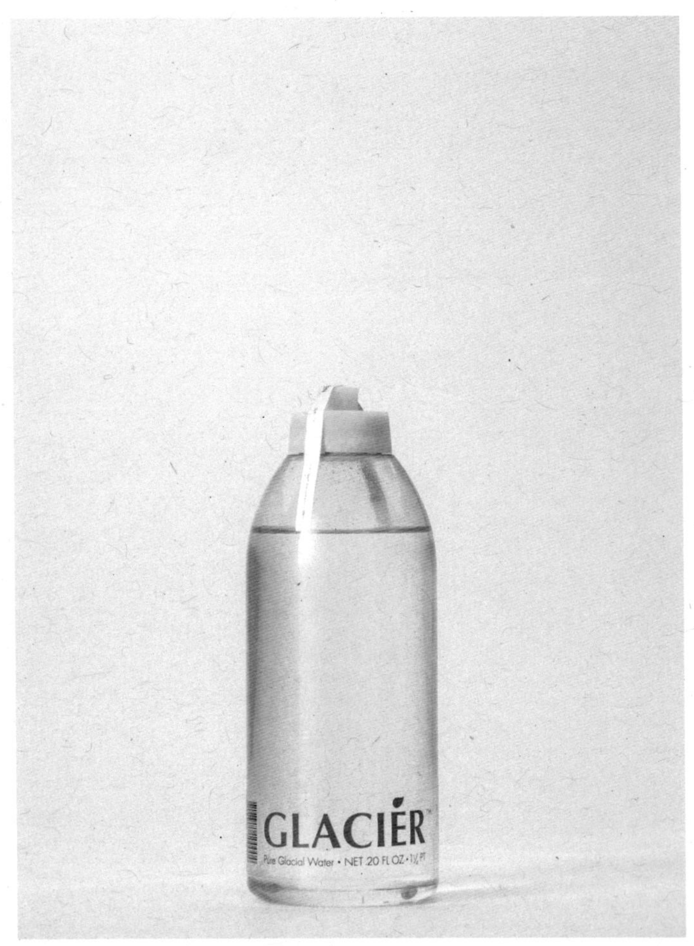

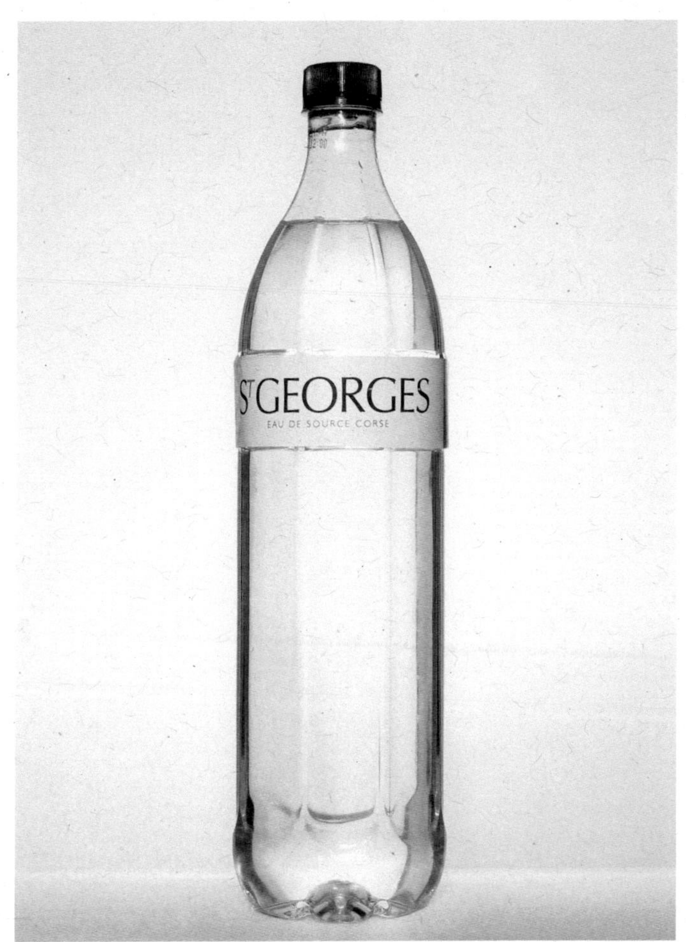

STGEORGES
EAU DE SOURCE CORSE

* "This is *Turdus Aonalaschkae Pallasii*, the hermit-thrush which I have heard in Quebec province. Chapman says (*Handbook of Birds of Eastern North America*) 'It is most at home in secluded woodland and thickety retreats... its notes are not remarkable for variety or volume, but in purity and sweetness of tone and exquisite modulation they are unequaled. 'Its water-drip-ping song' is justly celebrated" (Eliot's note.)

Here is no water but only rock
Rock and no water and the sandy road
The road winding above among the mountains
Which are mountains of rock without water
If there were water we should stop and drink
Amongst the rock one cannot stop or think
Sweat is dry and feet are in the sand
If there were only water amongst the rock
Dead mountain mouth of carious teeth that cannot spit
Here one can neither stand nor lie nor sit
There is not even silence in the mountains

But dry sterile thunder without rain
There is not even solitude in the mountains
But red sullen faces sneer and snarl
From doors of mudcracked houses
If there were water

And no rock
If there were rock
And also water
And water
A spring
A pool among the rock
If there were the sound of water only
Not the cicada
And dry grass singing
But sound of water over a rock
Where the hermit-thrush sings in the pine trees
Drip drop drip drop drop drop drop *
But there is no water

—T.S. ELIOT, *THE WASTE LAND*

"Flood of the Century": Manitoba sites during & after *Isaac Applebaum* Photographs of the 1997 Red River flood and the land revealed when the waters retreated.

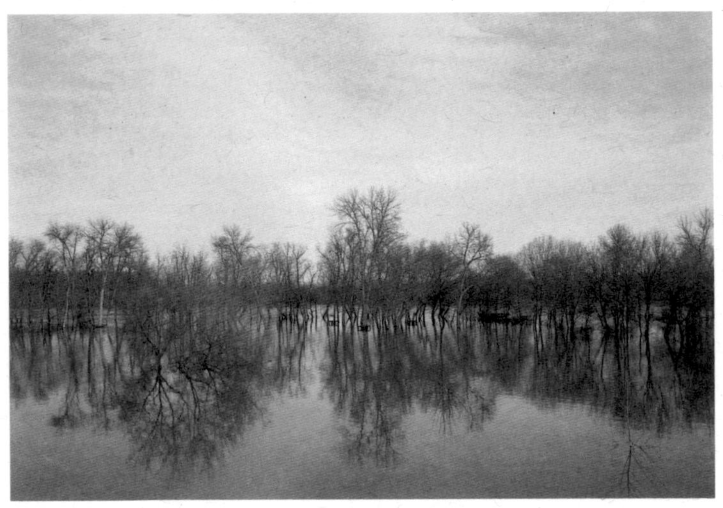

Outside of Selkirk; May & August 1997

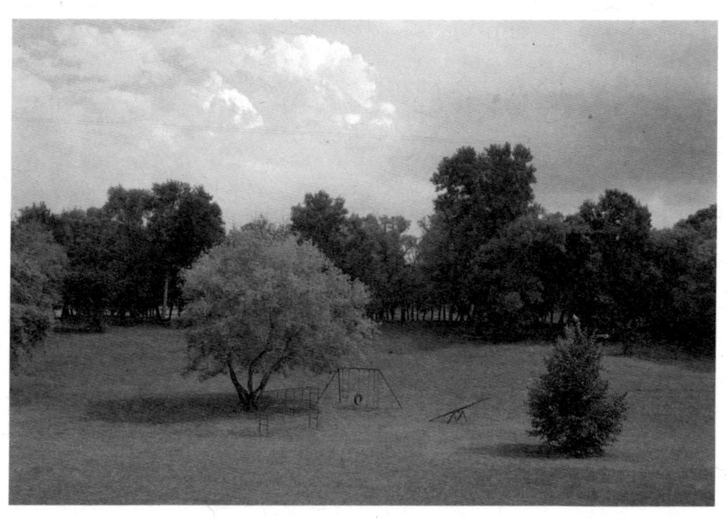

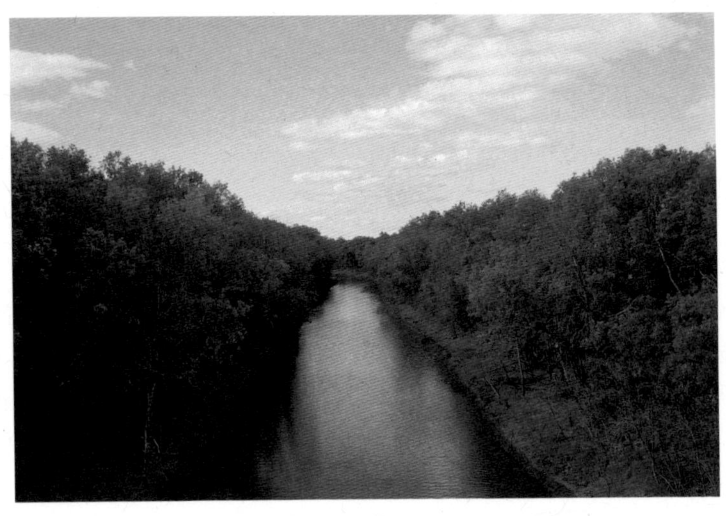

Rat River, just north of Ste. Agathe: August & May 1997

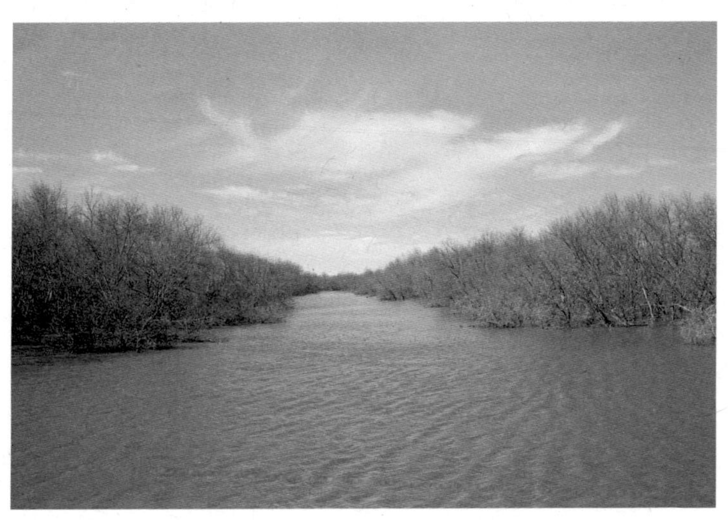

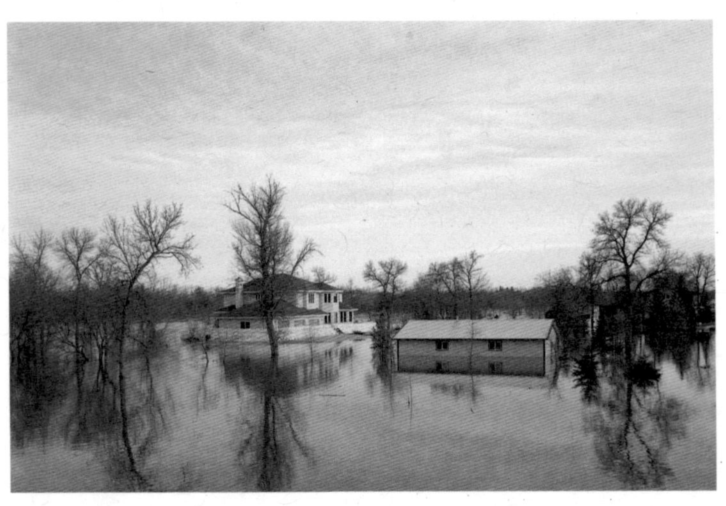

Outside of Selkirk; May & July 1997

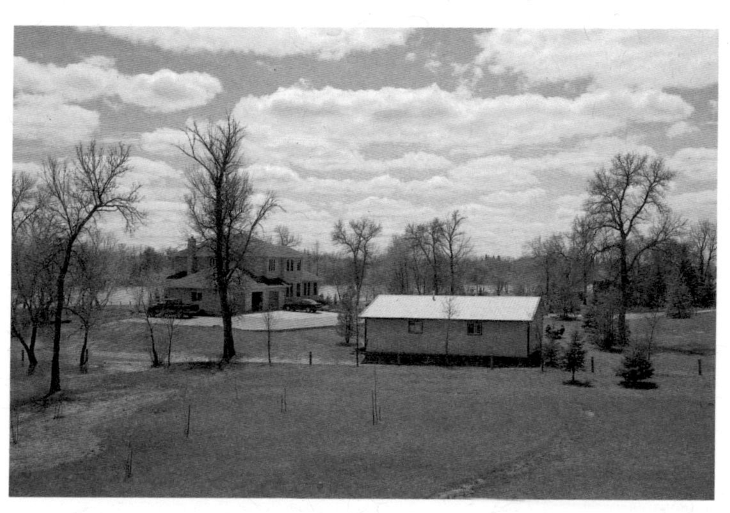

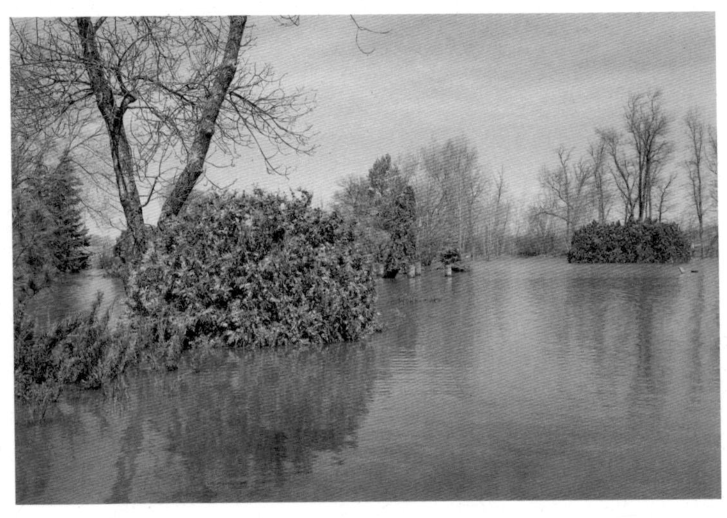

Between Winnipeg & Lockport; May & August 1997

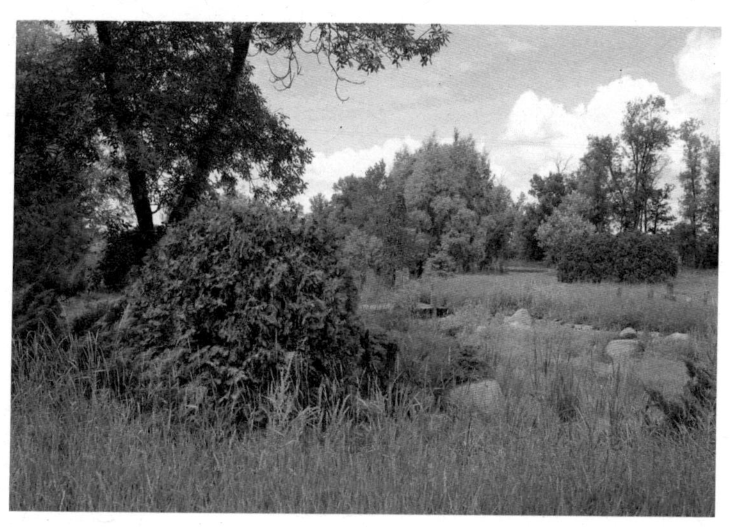

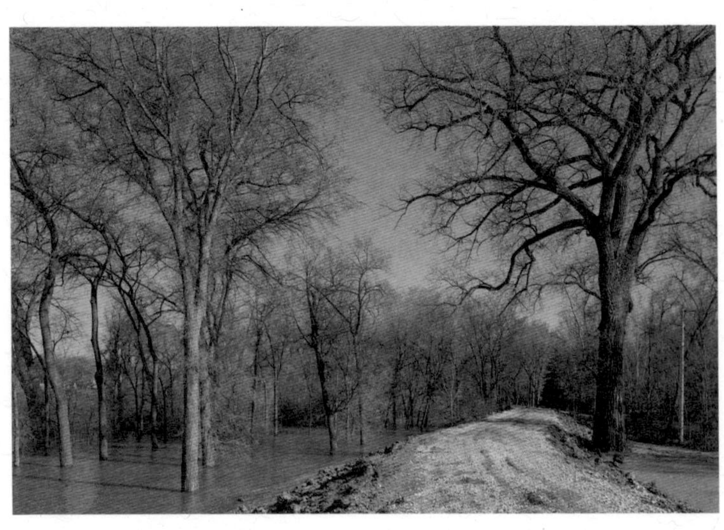

North End Winnipeg; May & August 1997

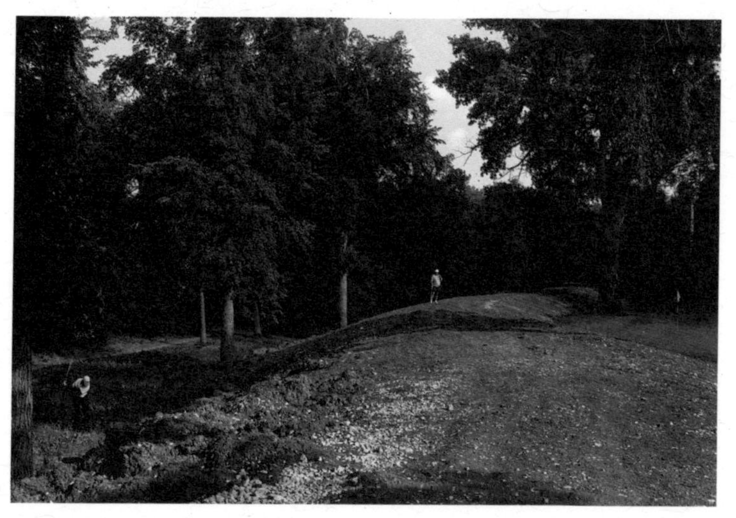

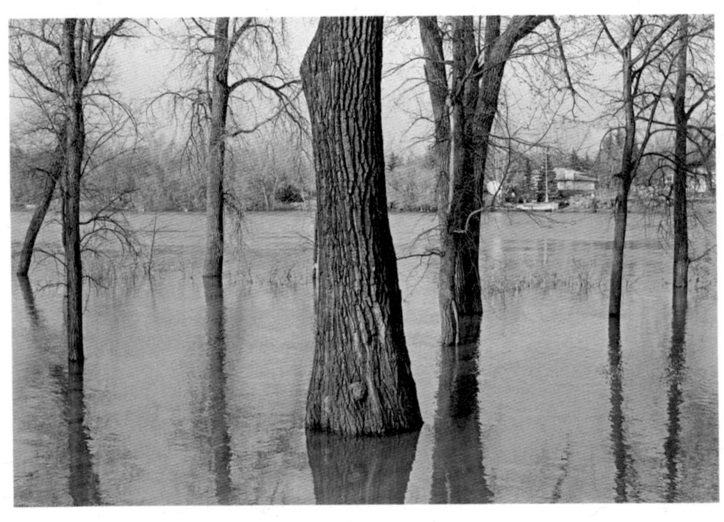

Kildonan Park, Winnipeg; May & July 1997

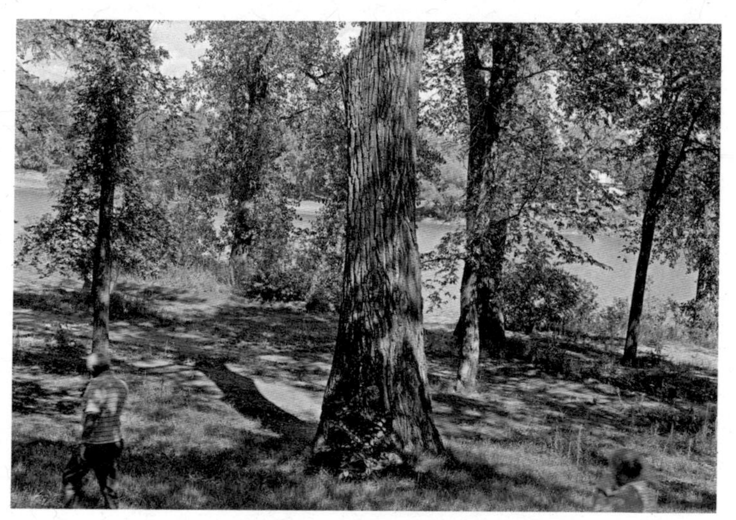

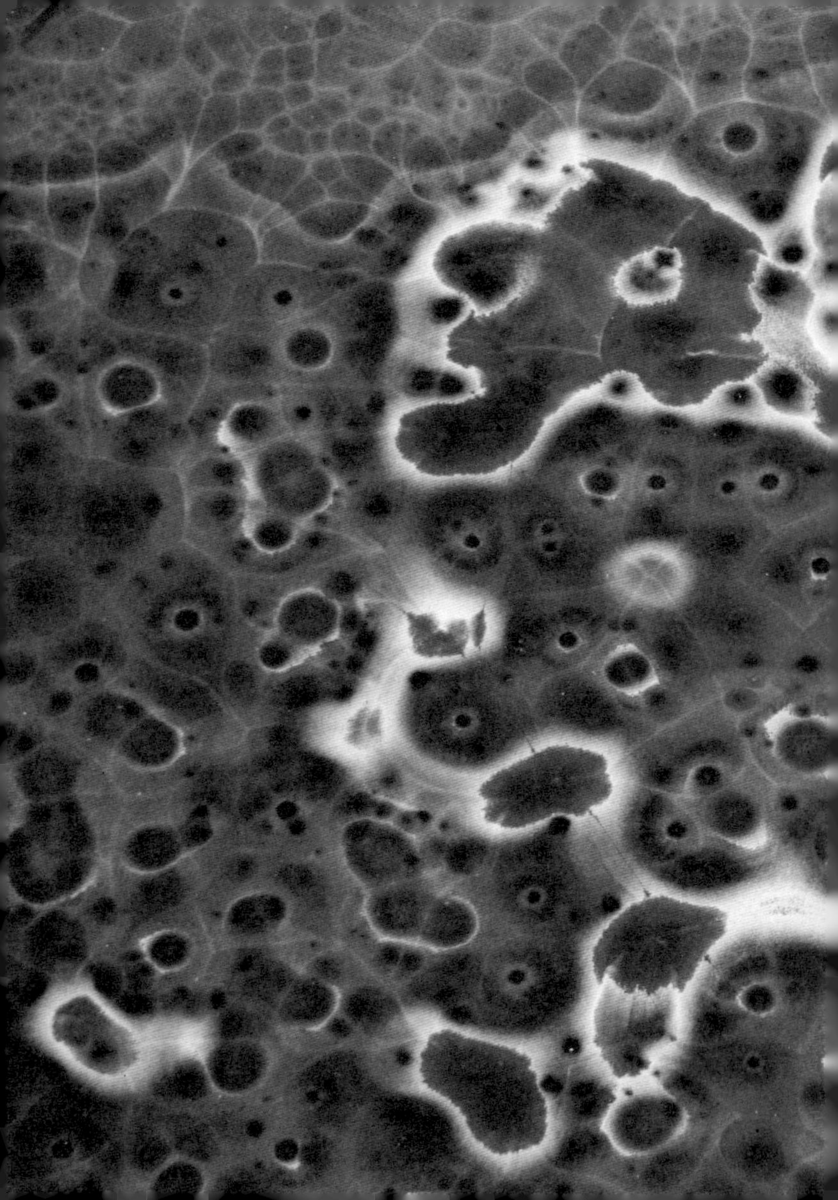

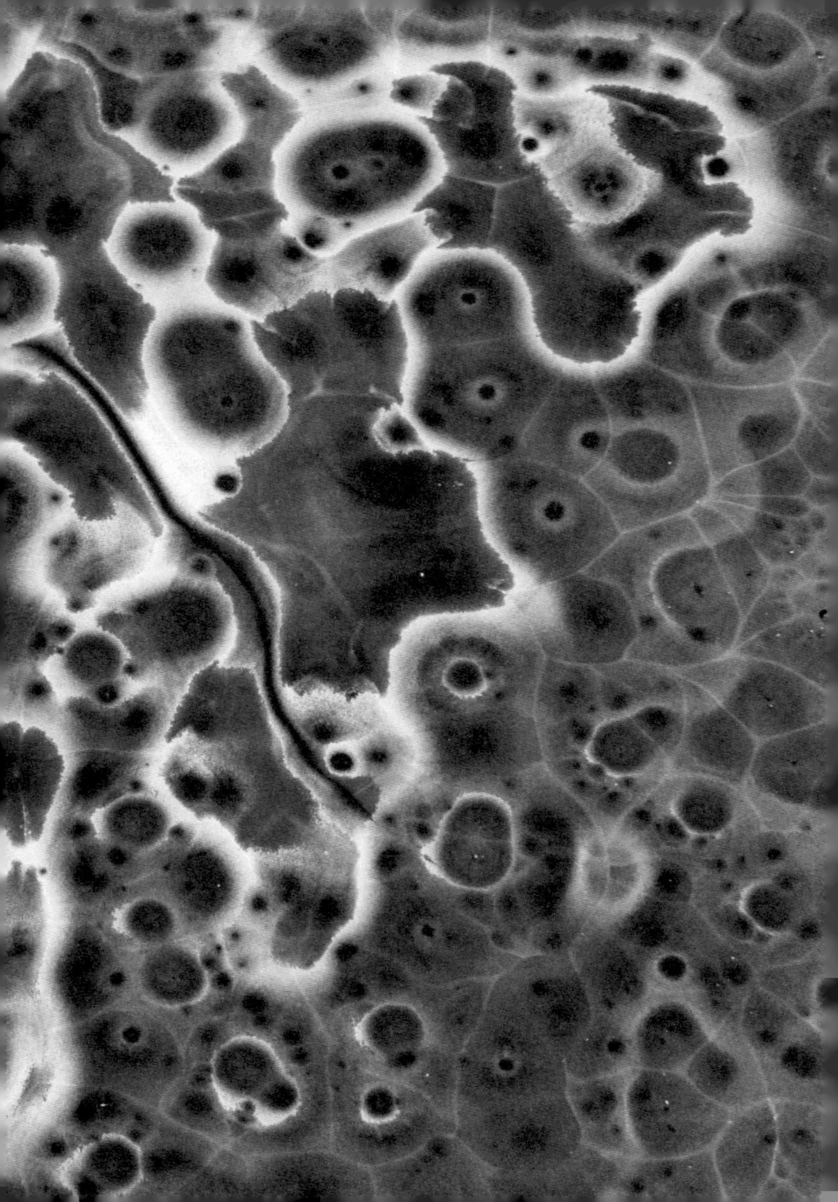

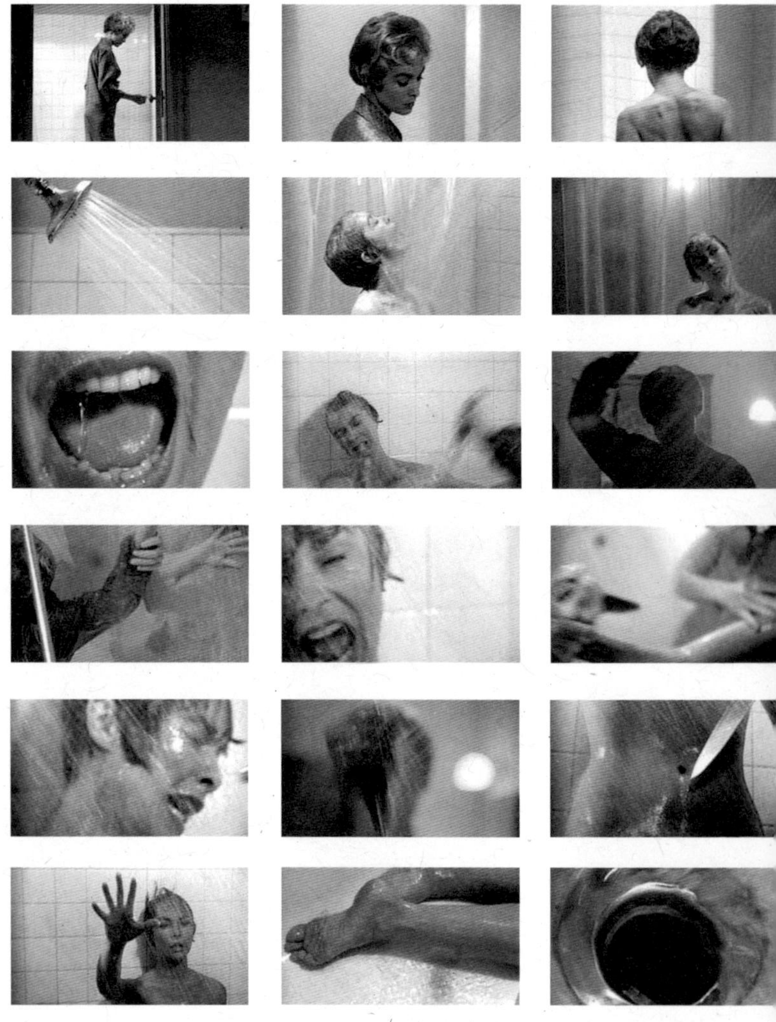

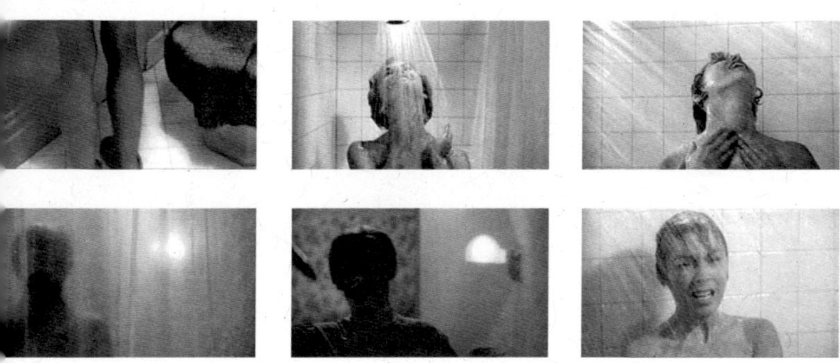

Psychodrama: 13 Variations: for eleven instruments, tape, and video *Melissa Grey*

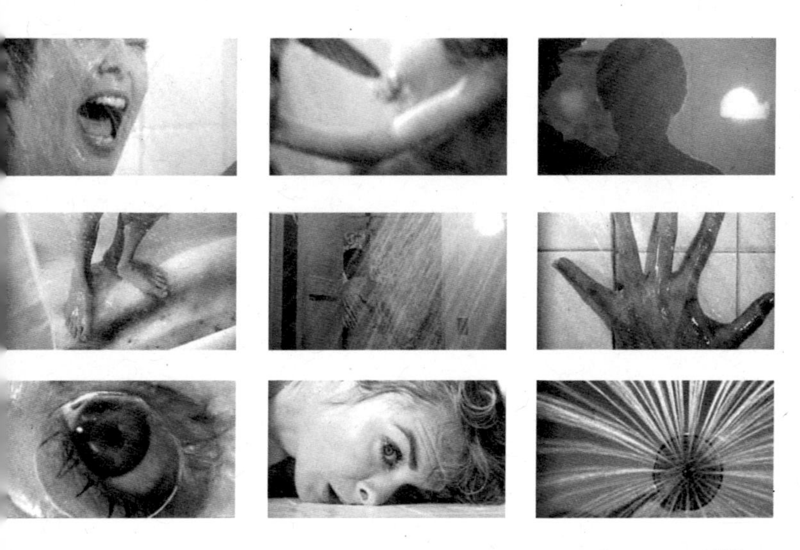

Variation No. 1

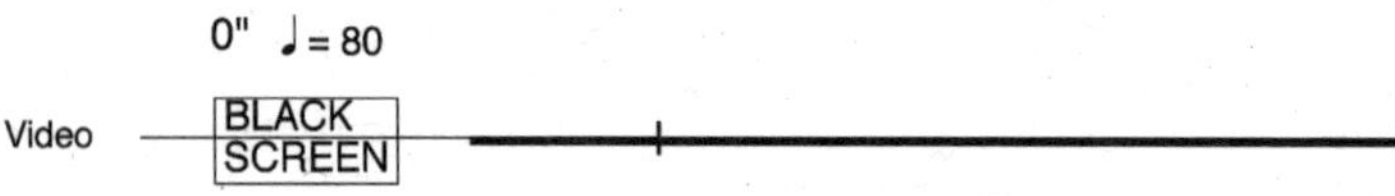

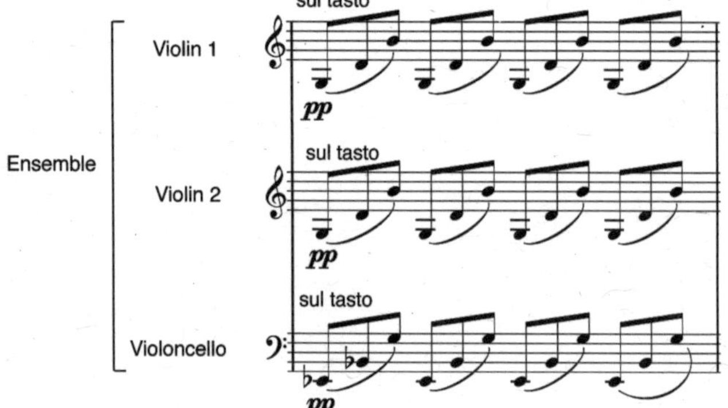

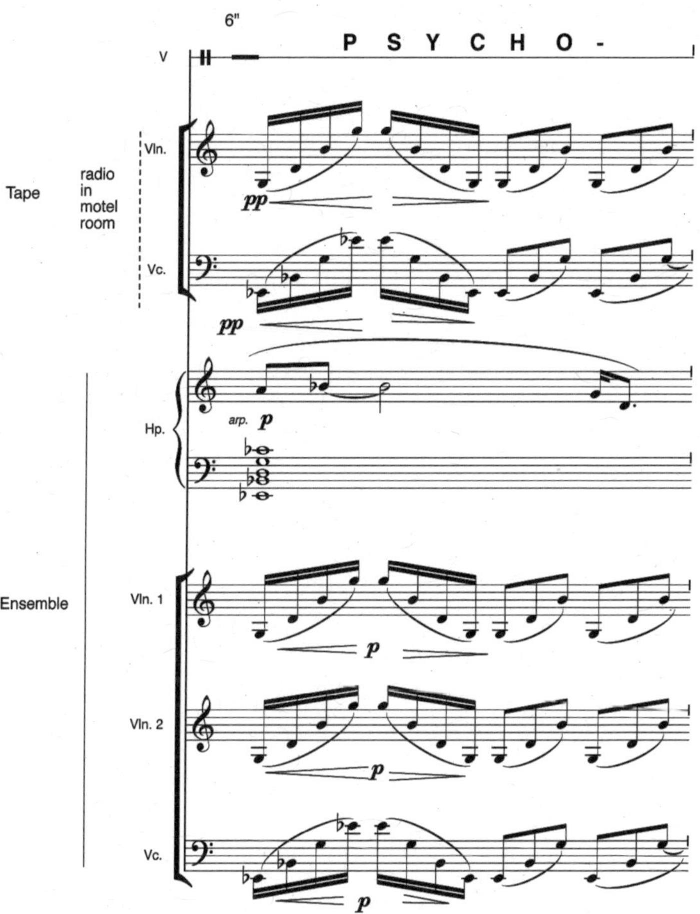
6"
P S Y C H O -
V
Vln.
Tape
radio
in
motel
room
Vc.
pp
arp. p
Hp.
Ensemble
Vln. 1
p
Vln. 2
Vc.

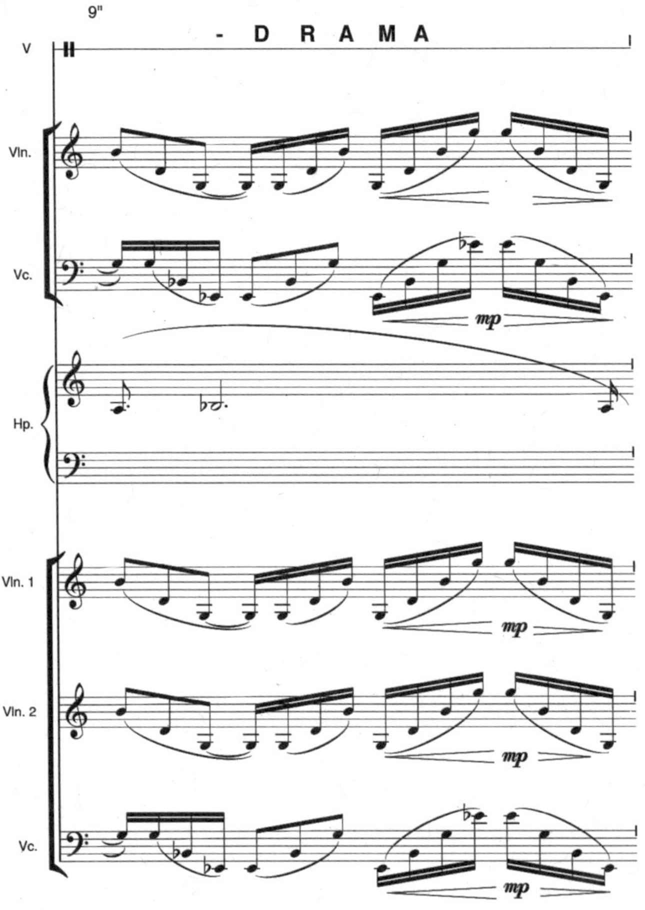
9"
- D R A M A
V
Vln.
Vc.
mp
Hp.
Vln. 1
mp
Vln. 2
mp
Vc.
mp

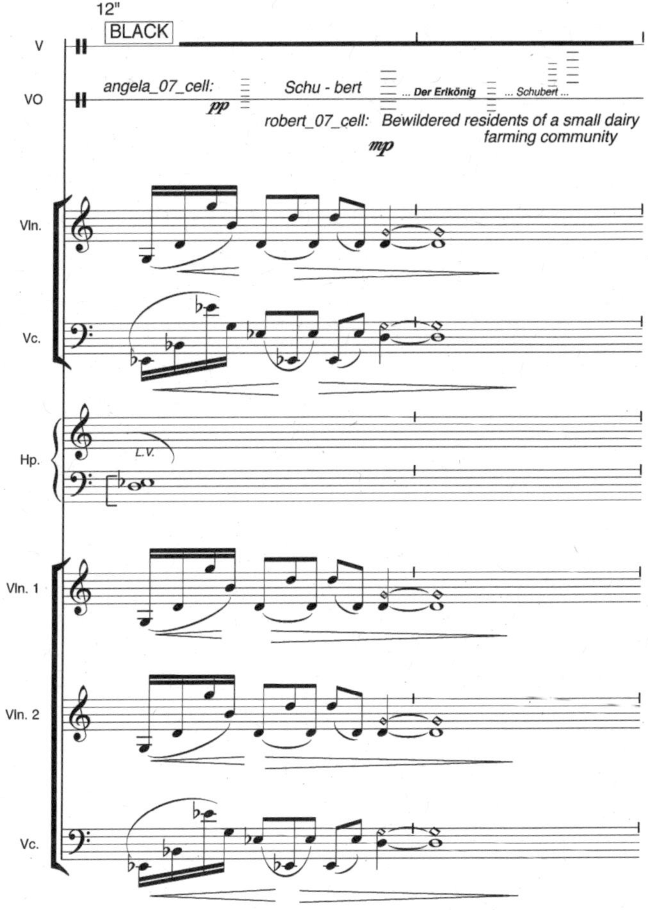
12"
BLACK
V
VO
angela_07_cell:
pp
Schu - bert
... Der Erlkönig
... Schubert ...
robert_07_cell: Bewildered residents of a small dairy farming community
mp
Vln.
Vc.
Hp.
L.V.
Vln. 1
Vln. 2
Vc.

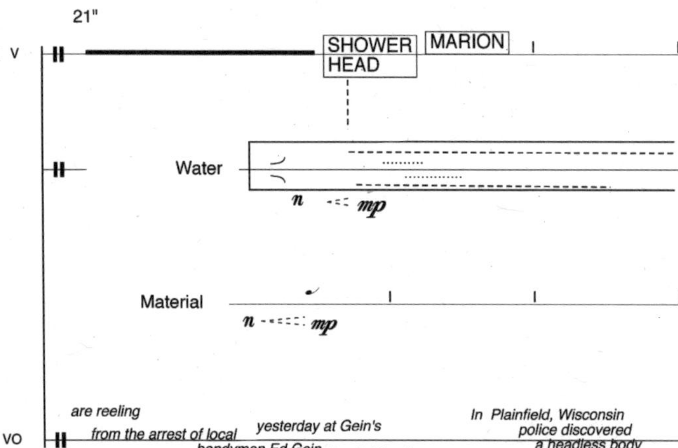

21"
V
SHOWER HEAD
MARION
Water
n
mp
Material
n
mp
VO
are reeling
from the arrest of local
handyman Ed Gein.
yesterday at Gein's
derelict farmhouse
In Plainfield, Wisconsin
police discovered
a headless body
and the remains of many others

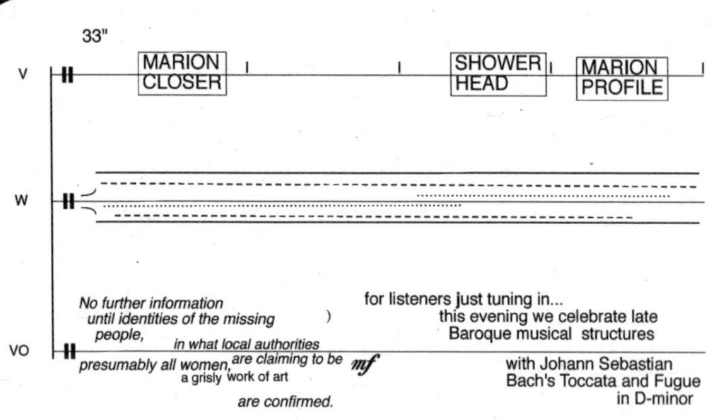

33"
V
MARION CLOSER
SHOWER HEAD
MARION PROFILE
W
VO
No further information
until identities of the missing
people,
presumably all women,
are confirmed.
in what local authorities
are claiming to be
a grisly work of art
mf
for listeners just tuning in...
this evening we celebrate late
Baroque musical structures
with Johann Sebastian
Bach's Toccata and Fugue
in D-minor

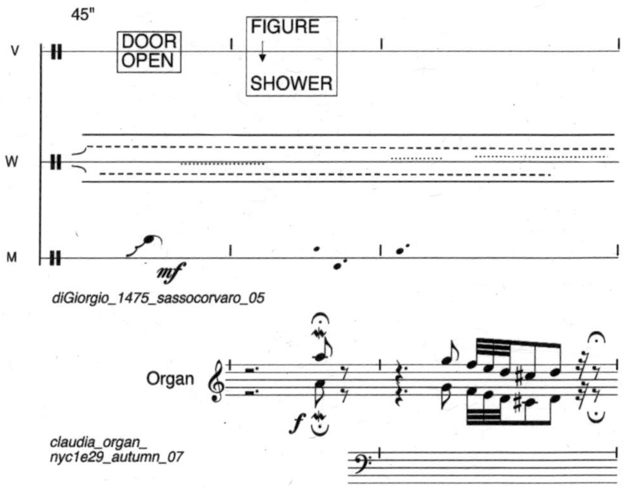
45"
V
DOOR
OPEN
FIGURE
SHOWER
W
M
mf
diGiorgio_1475_sassocorvaro_05
Organ
f
claudia_organ_
nyc1e29_autumn_07

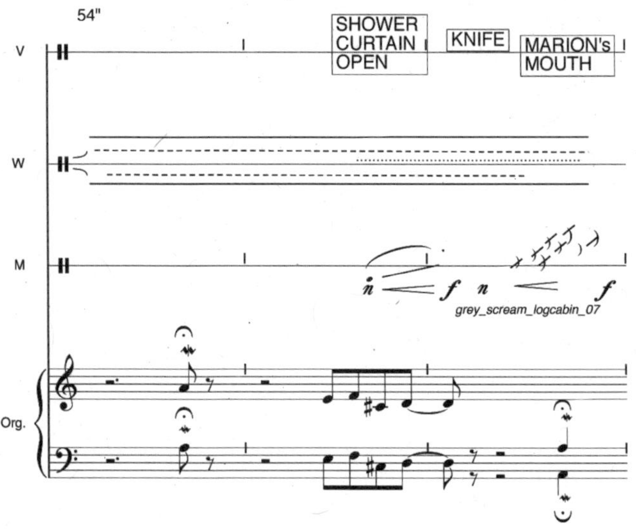
54"
V
SHOWER
CURTAIN
OPEN
KNIFE
MARION's
MOUTH
W
M
grey_scream_logcabin_07
Org.

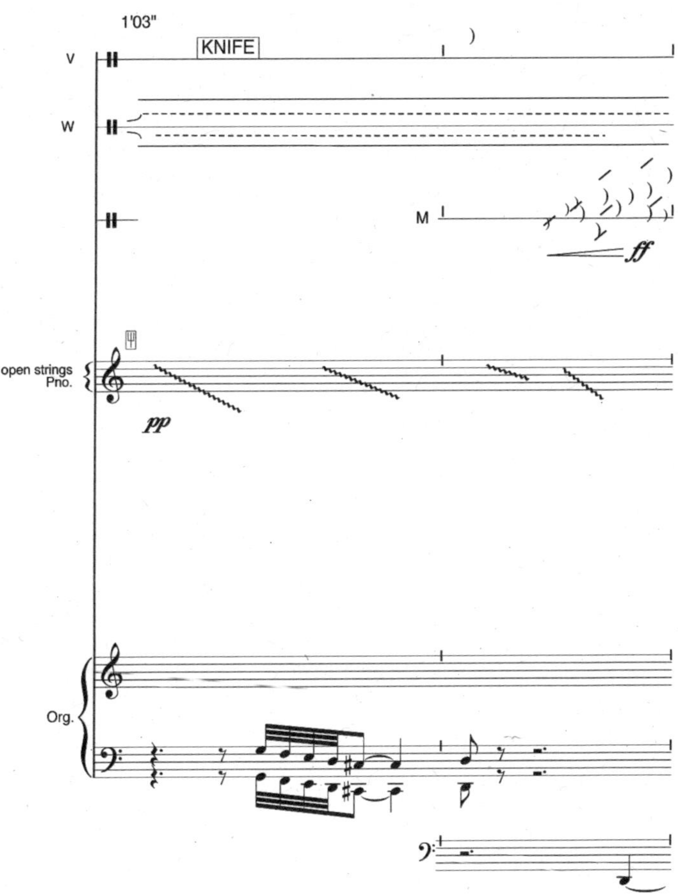
1'03"
KNIFE
V
W
M
ff
open strings
Pno.
pp
Org.

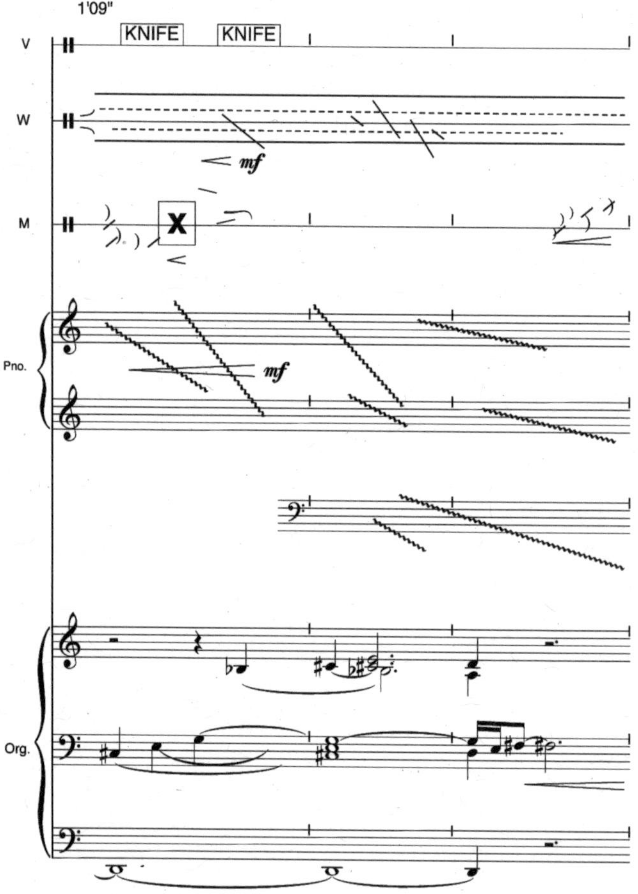
1'09"
KNIFE
KNIFE
V
W
mf
M
X
Pno.
mf
Org.

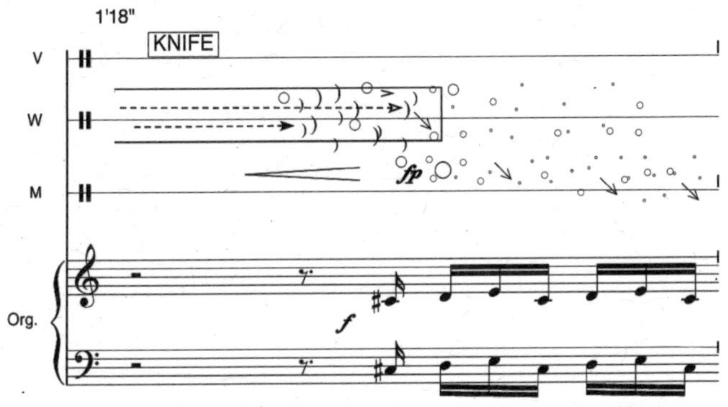

1'18"
KNIFE
V
W
M
fp
f
Org.

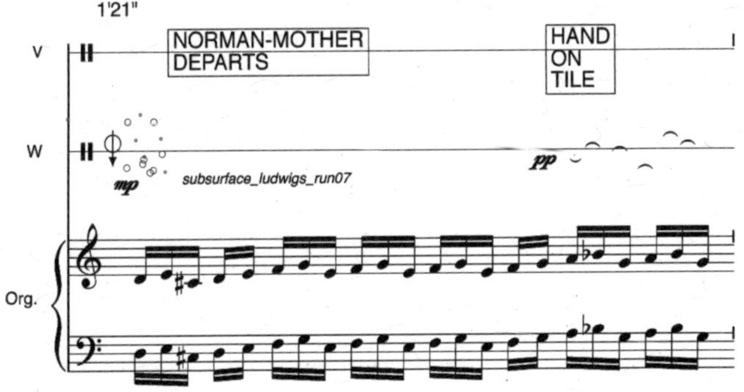

1'21"
NORMAN-MOTHER
DEPARTS
HAND
ON
TILE
V
W
mp
subsurface_ludwigs_run07
pp
Org.

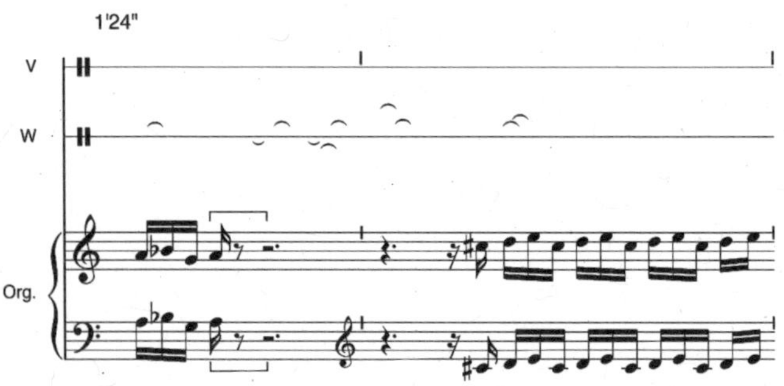
1'24"
V
W
Org.

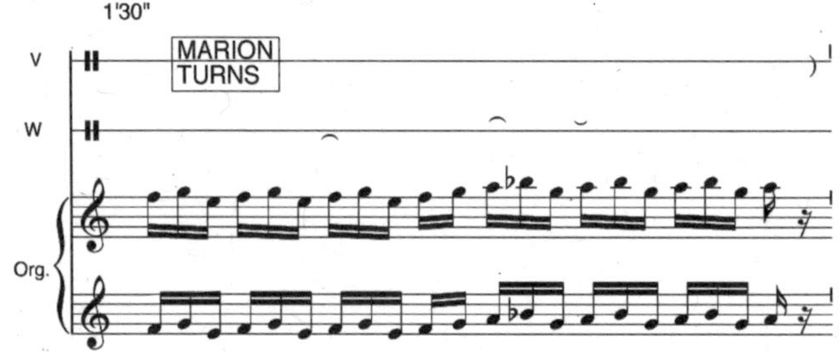
1'30"
MARION
TURNS
V
W
Org.

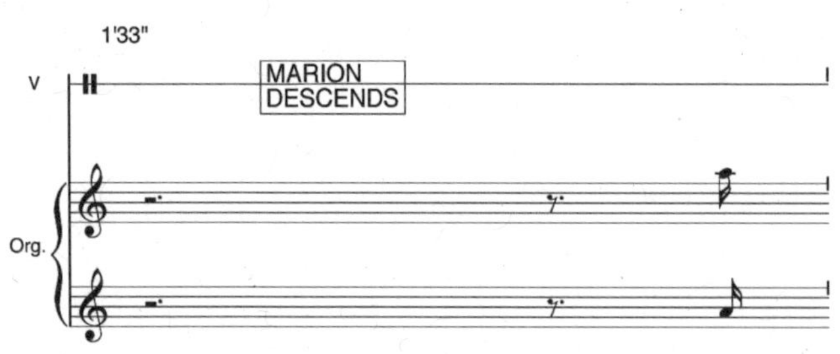
1'33"
MARION
DESCENDS
V
Org.

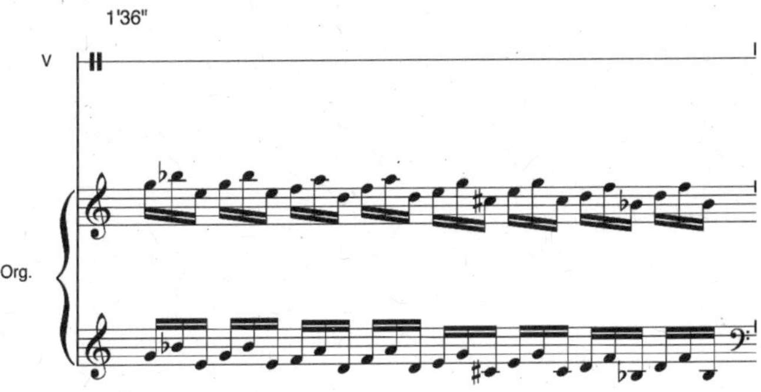
1'36"
V
Org.

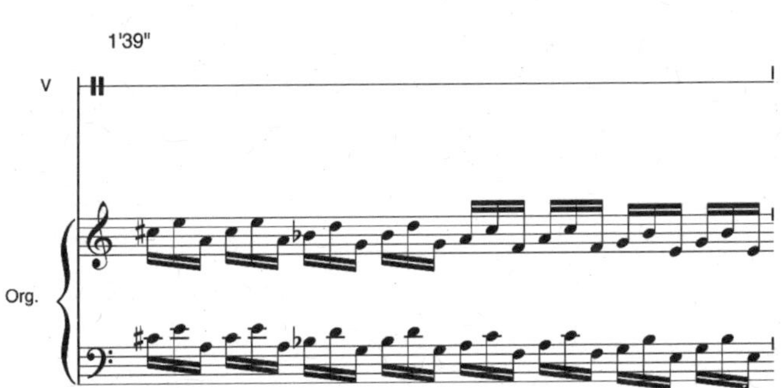
1'39"
V
Org.

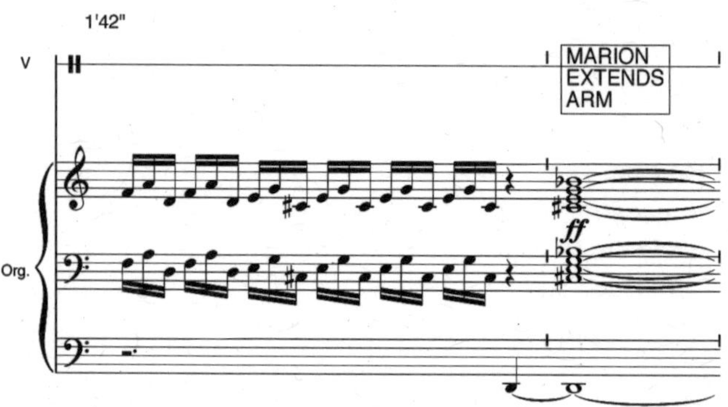
1'42"
V
MARION
EXTENDS
ARM
ff
Org.

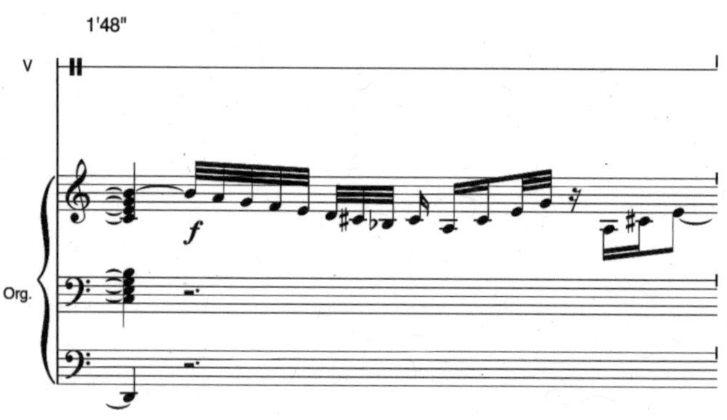
1'48"
V
f
Org.

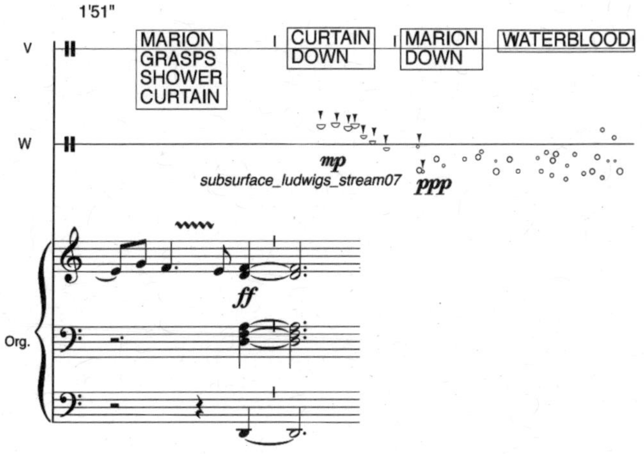
1'51"
V
MARION
GRASPS
SHOWER
CURTAIN
CURTAIN
DOWN
MARION
DOWN
WATERBLOOD
W
mp
subsurface_ludwigs_stream07
ppp
ff
Org.

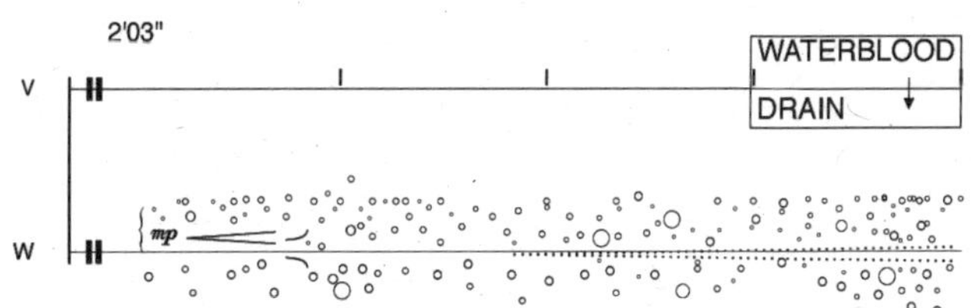
2'03"
WATERBLOOD
V
DRAIN
mp
W

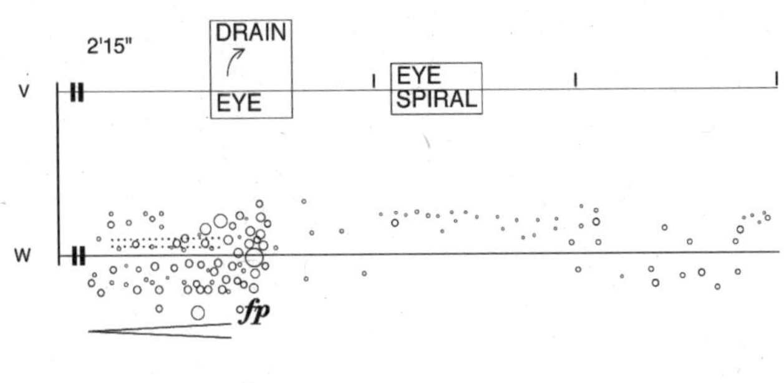
2'15"
DRAIN
EYE
EYE
SPIRAL
V
W
fp

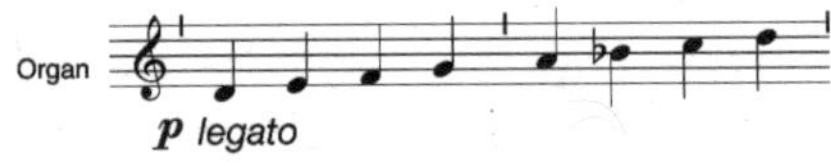
Organ
p legato

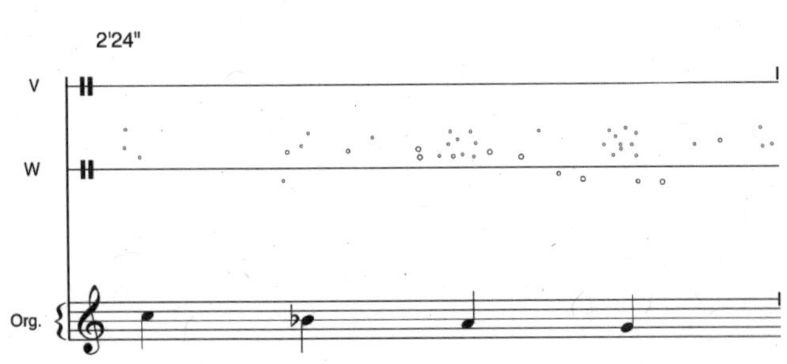
2'24"
V
W
Org.

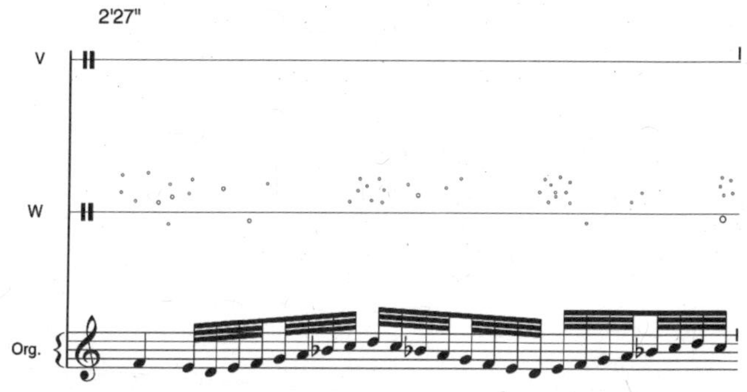
2'27"
V
W
Org.

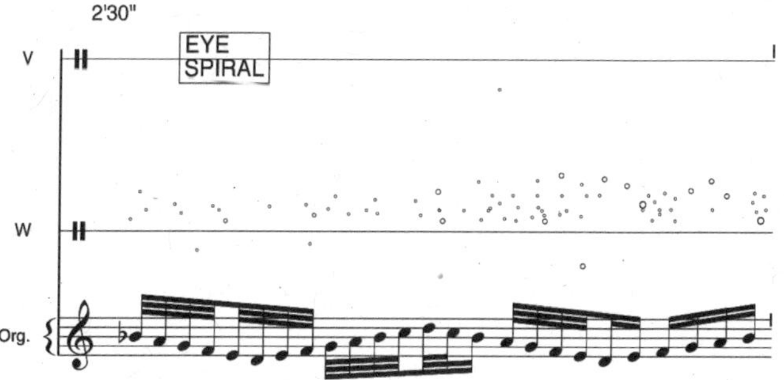
2'30"
EYE
SPIRAL
V
W
Org.

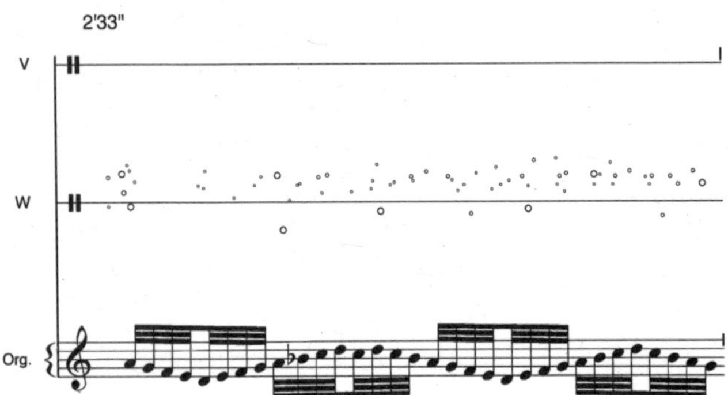

2'33"
V
W
Org.

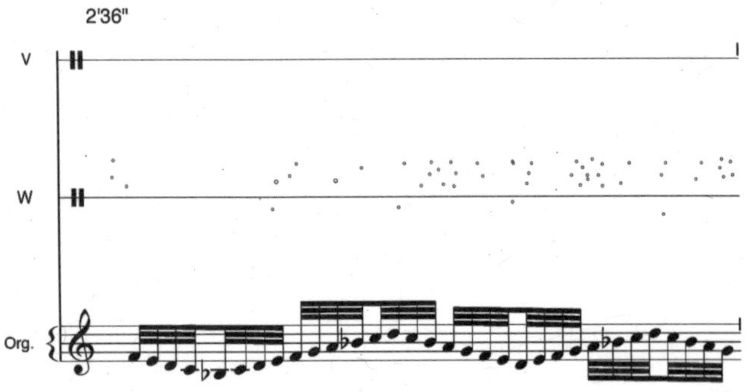

2'36"
V
W
Org.

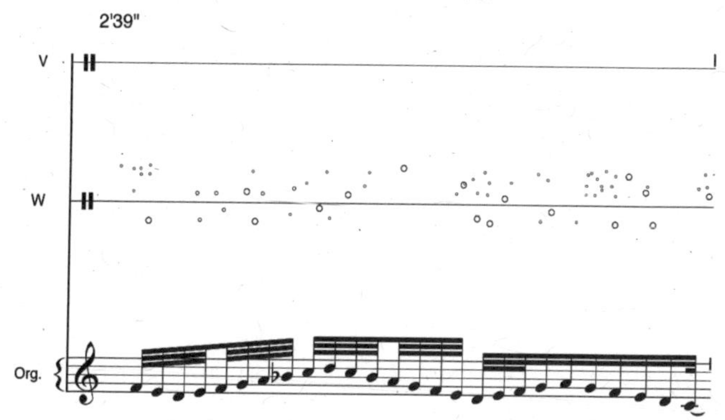
2'39"
V
W
Org.

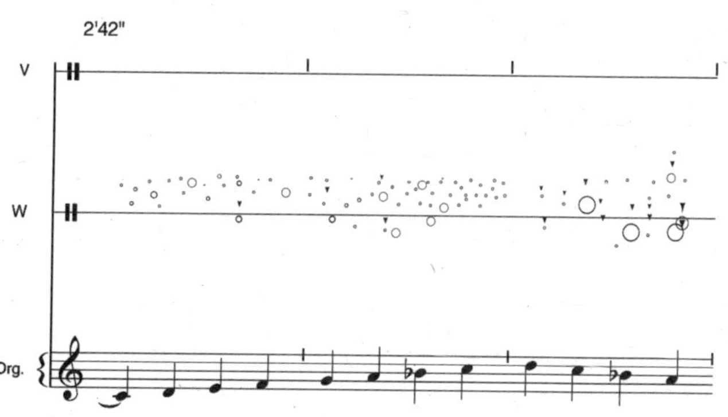
2'42"
V
W
Org.

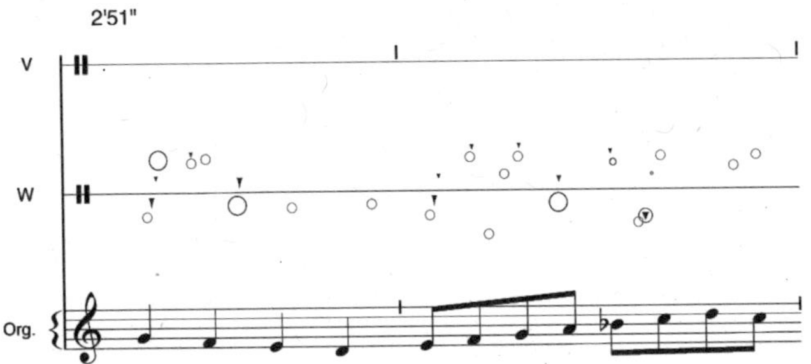
2'51"
V
W
Org.

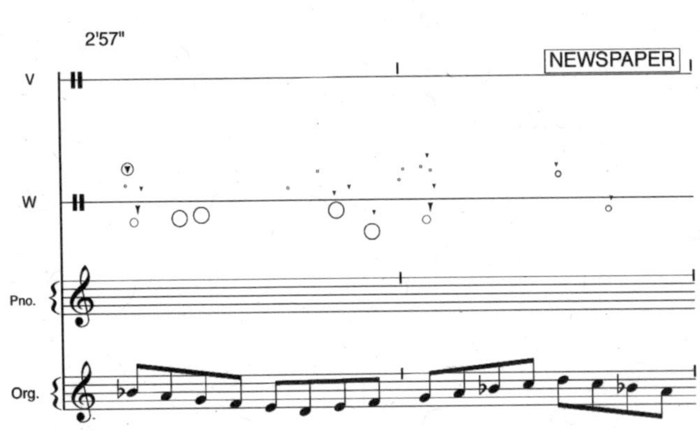
2'57"
NEWSPAPER
V
W
Pno.
Org.

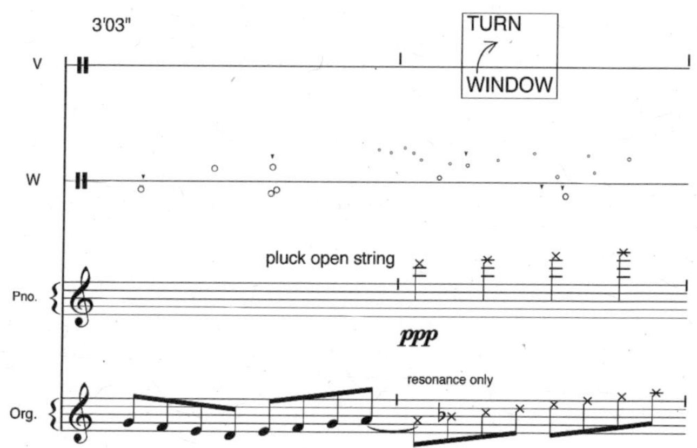
3'03"
V
TURN
WINDOW
W
pluck open string
Pno.
ppp
resonance only
Org.

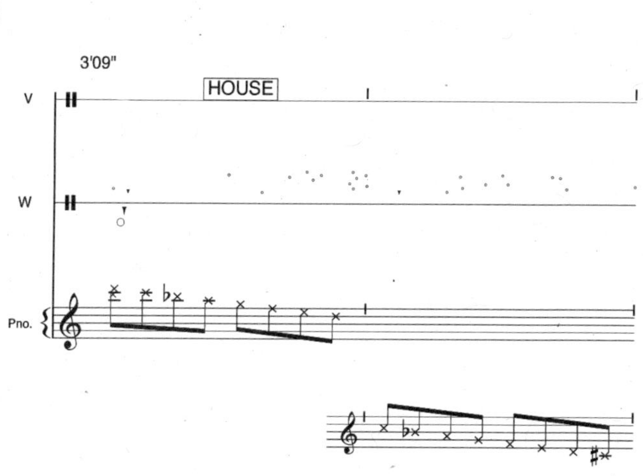
3'09"
HOUSE
V
W
Pno.

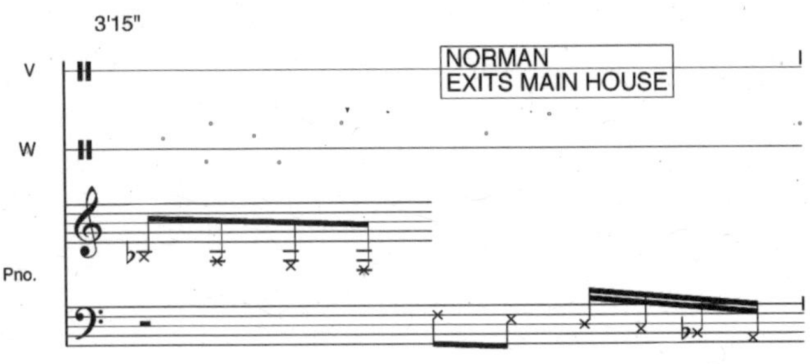
3'15"
V
W
Pno.
NORMAN
EXITS MAIN HOUSE

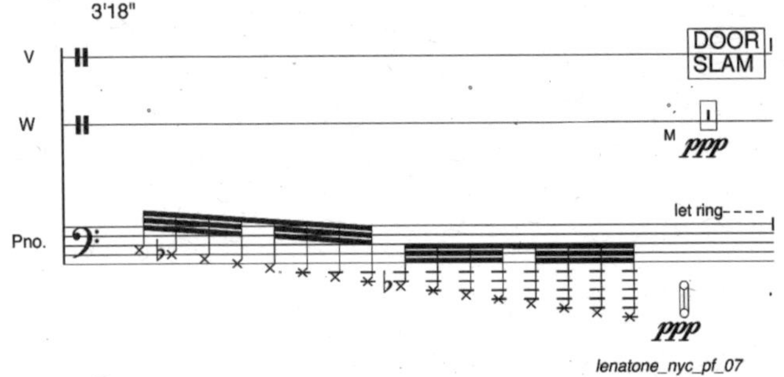
3'18"
V
W
Pno.
DOOR
SLAM
M
ppp
let ring
ppp
lenatone_nyc_pf_07

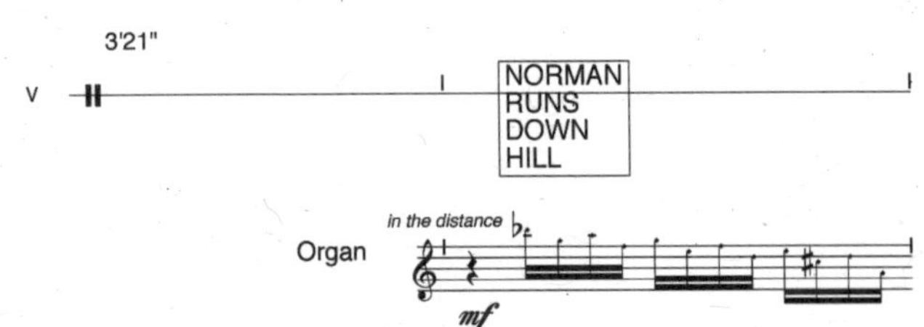
3'21"
V
NORMAN
RUNS
DOWN
HILL
in the distance
Organ
mf

3'27"

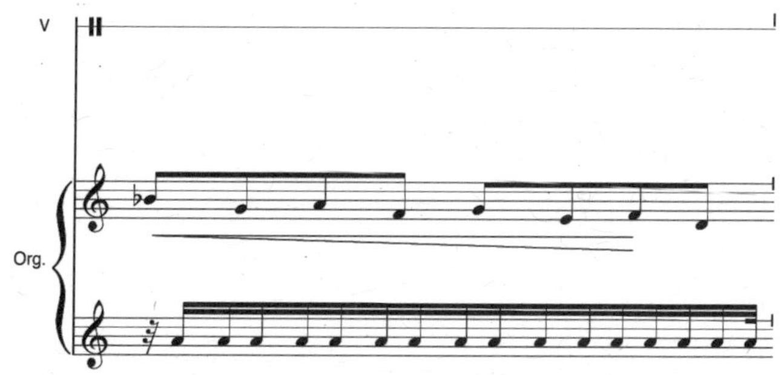
V
Org.

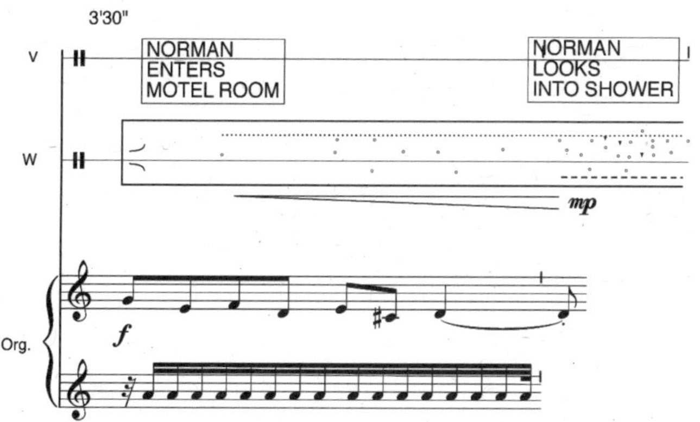
3'30"
V
NORMAN
ENTERS
MOTEL ROOM
NORMAN
LOOKS
INTO SHOWER
W
mp
f
Org.

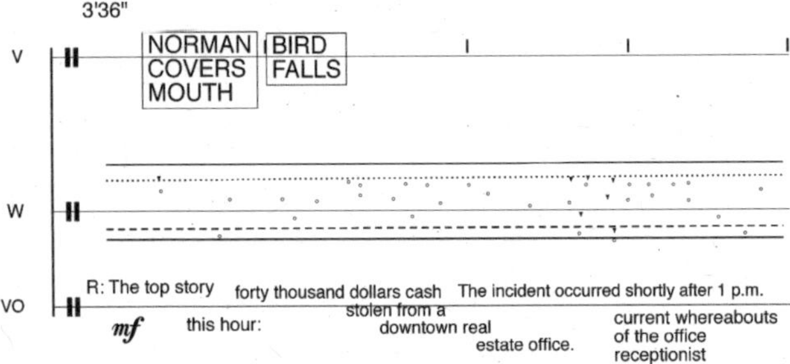
3'36"
V
NORMAN COVERS MOUTH
BIRD FALLS
W
VO
R: The top story
this hour:
forty thousand dollars cash
stolen from a
downtown real
estate office.
The incident occurred shortly after 1 p.m.
current whereabouts
of the office
receptionist
mf

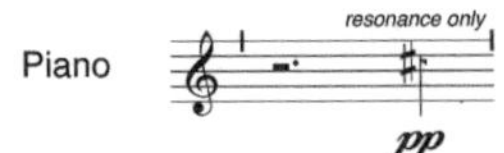
Piano
resonance only
pp

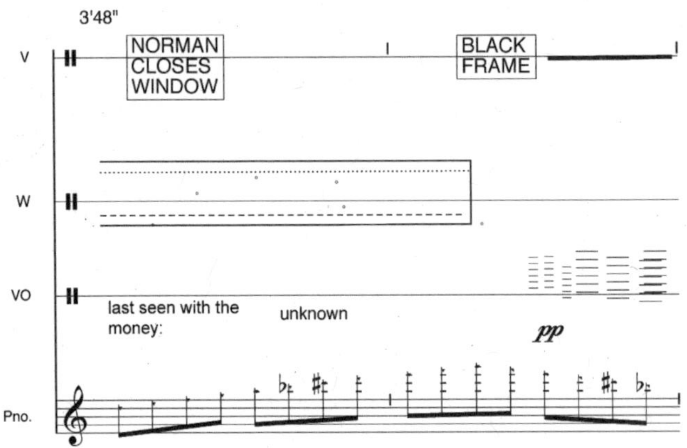
3'48"
V
NORMAN CLOSES WINDOW
BLACK FRAME
W
VO
last seen with the
money:
unknown
pp
Pno.

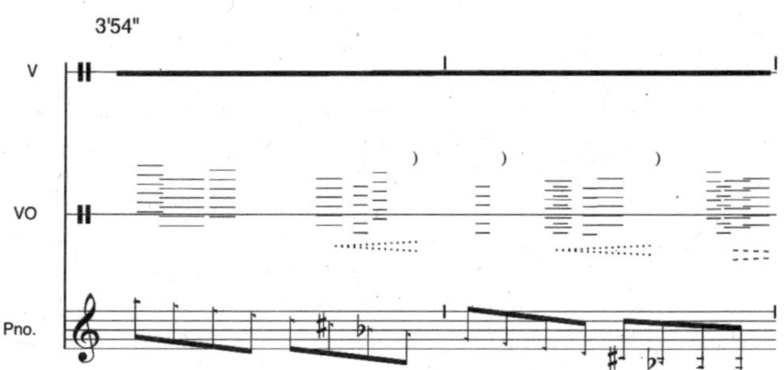
3'54"
V
VO
Pno.

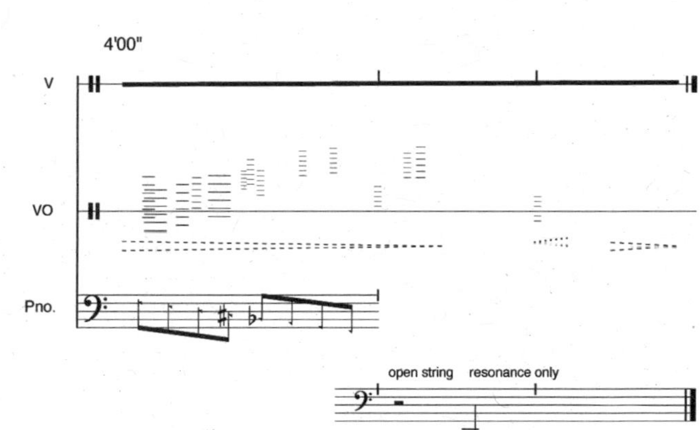
4'00"
V
VO
Pno.
open string
resonance only
4'12

WATER/FRONT *Nina-Marie Lister*

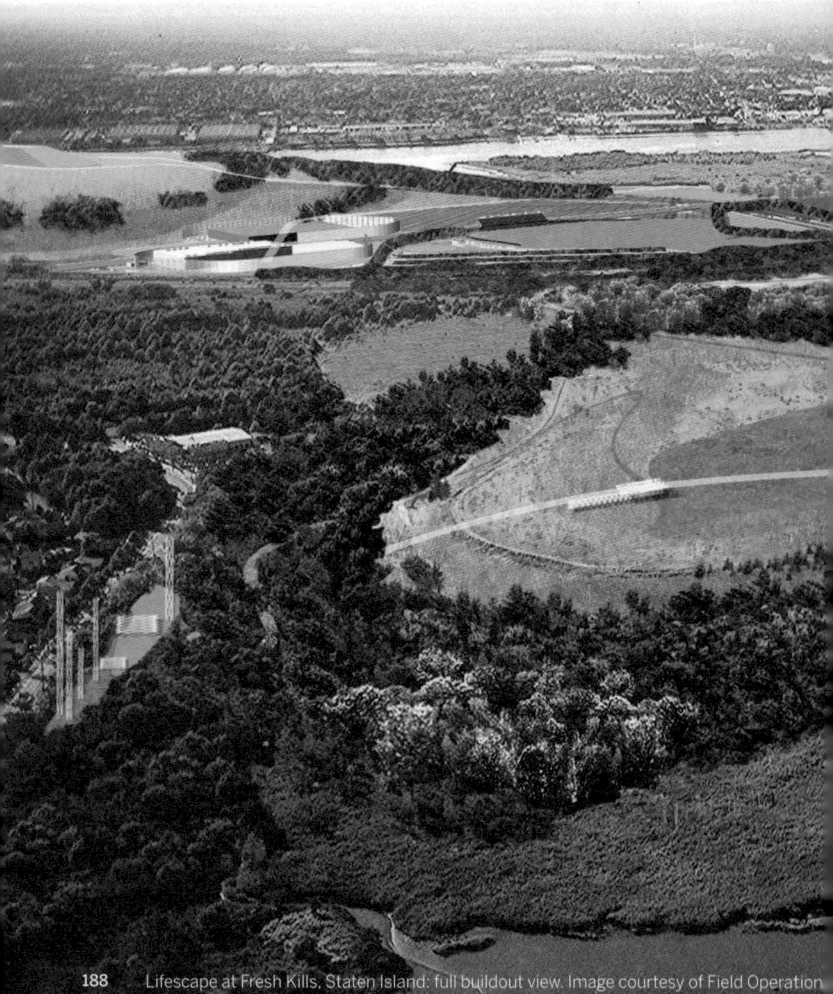

 Lifescape at Fresh Kills, Staten Island: full buildout view. Image courtesy of Field Operation

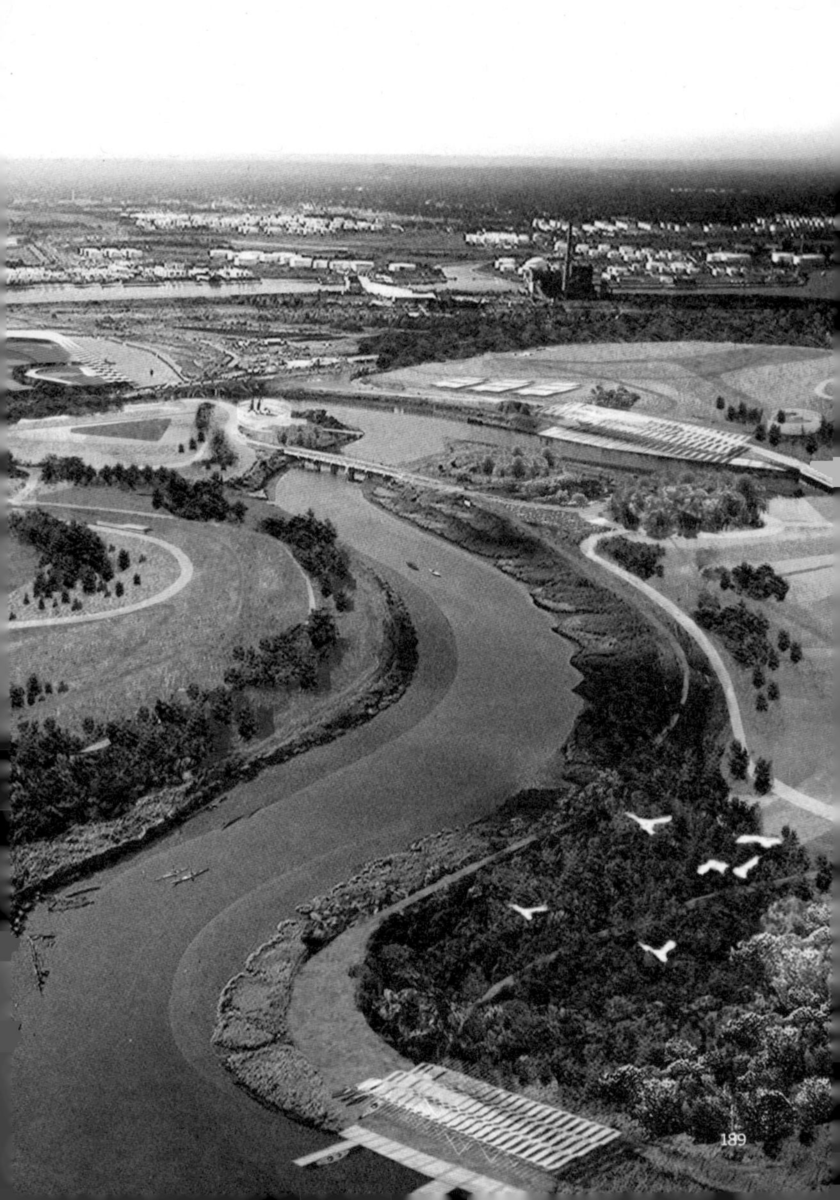

Virtually all of the world's significant cities embrace the water: Rome on the Tiber, Paris on the Seine, London on the Thames, New York on the Hudson, Hong Kong on the Pearl Delta, and many more. In these cities, the water's edge is a messy confluence of rich and poor—beach houses and slums—held in dynamic tension between luxury and filth, salvation and rejection, taking and dumping. Many of these urban waterfronts have declined from their former glory into a state of decay. But around the world, cities are taking a fresh look at their dockyards and beaches and seeking to remediate their relationship with the water's edge.

In the past, waterfront lands were favored sites for industry. Factories were built close to the point of shipping, where raw materials and commodities could conveniently be delivered by nearby road, rail, or sea. Refineries, mills, rendering and chemical plants clustered at the transit nodes where goods were imported and exported. As cities and economies grew, so too did the byproducts of progress. Effluents accumulated, marine life died, and the waters became sick; flooding, pollution, and sedimentation choked and slowed formerly clean-flowing waters. By the height of the industrial era, only the poorest folk lived near the water; harbors and rivers were associated with dirt and disease. As water transport declined and freight was increasingly moved by road, the docklands lost their commercial value. Businesses and residents moved away, and the once sacred water's edge was all but abandoned. Contaminated, desolate, and left to decay, waterfronts the world over became little more than dirty backyards that the cities had quite literally turned their backs on and forgotten.

Yet as city-regions grew and their economies began to shift from manufacturing to a service and knowledge base, governments began to reassess their underused port lands. The forces of a changing labor market, expanding suburbs, shrinking downtowns, and deindustrialization combined to open new opportunities for underused waterfront areas; even contaminated and derelict post-industrial sites began to gain in value. Port lands, harbors, and riverfronts are now being considered for new uses that have the potential to generate income and revitalize the urban core. Many cities are now committed to the costly and complicated processes of cleanup in the service of long-term economic development. A new urban engagement with the water's edge is emerging. Early examples in London's Docklands and New York's Battery Park City have led to a groundswell of revitalization efforts worldwide, as city-regions are rediscovering and confronting the water. The water's edge is a front, contested territory at the doorstep of the post-modern metropolis, where the conflicts around water—from consumption to conservation—in brownfields, and greenfields are being played out as new projects are planned and set in motion.

Hamburg, Sydney, Seoul, Mumbai, New York, and Toronto are but a few examples of major cosmopolitan cities undertaking significant waterfront projects. Major waterside parklands are being planned, arts and cultural centers built; entirely new residential and commercial communities are emerging. Virtually all of these urban initiatives are centered on contaminated and brownfield sites, where in many cases the cost of cleanup was seen as too high if left exclusively to the public sector or to the now-defunct industries that caused the problem. Now, more creative ways are being found to increase the tax base and reboot flagging local economies. Government agencies or public–private sector partnerships are selling or leasing derelict waterfront lands to private developers, usually multinational consortia, who will design and build new mixed-use residential-commercial communities. The selection process for the urban design team typically includes an international competition intended to increase the project's profile, attract investors, and leverage political support. For example, Sydney's revitalization of East Darling Harbor effectively used a competition to rebrand the area as "Barangaroo Park," and the short-listed design teams competed for the chance to develop a 22 hectare office park and mixed-use residential community, offering panoramic views of the famed Sydney Harbor. The winning plan included a 14 hectare park on the foreshore in which native vegetation would be returned to an area that was for many years a major shipping and container port.

A secondary but essential goal of these projects is to remediate or clean up the residual—in some cases extensive—contamination of the project sites. As a result, many waterfront redevelopments have a prominent sustainability component and "green" building design. Some of the more progressive renewal plans, such as Stockholm's Hammarby-Sjöstad, Hamburg's HafenCity and the Melbourne Docklands, specify detailed design criteria to achieve sustainable communities. These may combine market-based and social housing, alternative energy generation, multi-modal transit, community centers, parklands, and mixed-use commercial-residential neighborhoods with generous green space.

But another shift is occurring, one that goes beyond green building initiatives and pedestrian-friendly neighbourhoods. In a new wave of post-industrial revitalization, New York, Mumbai, and Toronto each are pioneering innovative approaches to site regeneration that call on designers to consider land and water in synergy. In such a coupled system, dynamic processes, performance, and adaptation through time are the most important goals, replacing outmoded emphases on static structures and the illusion of control through "preservation."

In these projects, designers focus on strategies that harness the complex ecologies of culture and nature at the water's edge. Notable projects using this ecological approach include the urban design firm Field Operations' Lifescape, the master plan for Fresh Kills Park on New York's Staten Island; Anuradha Mathur and Dilip da Cunha's SOAK: Mumbai in an Estuary, a speculative recalibration of that city's waterfront; and Stoss Landscape Urbanism's RIVER+CITY+LIFE, a proposed plan for Toronto's waterfront. All three projects engage culture, nature, and time in ways that challenge our ideas of what it means to live, work, and play at the water's edge.

At 937 hectares, Fresh Kills on Staten Island, New York, is one of the world's largest waterfront landfills, a site that recently became both grave and monument when the remains of the World Trade Center were added to its landscape. With Lifescape, the designers project a radical transformation of Fresh Kills into an urban park. Yet the challenges of the site demand a subtle, long-term approach to planning: the designers could not simply drop a conventional park on top of a trash heap, nor could they impose a didactic landscape lecture on the legacy of waste and the perils of consumption. In response to the site's ecological, cultural, and economic complexities, the design team chose to approach Fresh Kills as a story of life unfolding. As Robert Sullivan observed in the November 23, 2008, edition of the *New York Times Magazine*, "they would make the old dump a part of the new park, by acknowledging it, reclaiming it, recycling it on behalf of a modern metropolis." Consequently, the team's vision for Fresh Kills is more about process and performance than an arrangement of objects. The plan projects a decades-long succession of development stages, focused first on the trash mounds, then on the fields and prairies left as the mounds settle, followed by public open spaces, and finally as a programmable surface for events. In the short term, the plan for Fresh Kills keeps the impressive height of the trash mounds, and provides for the area to have restored wetlands, re-created tidal flats native to the Atlantic shore, and a variety of other native and designed ecologies

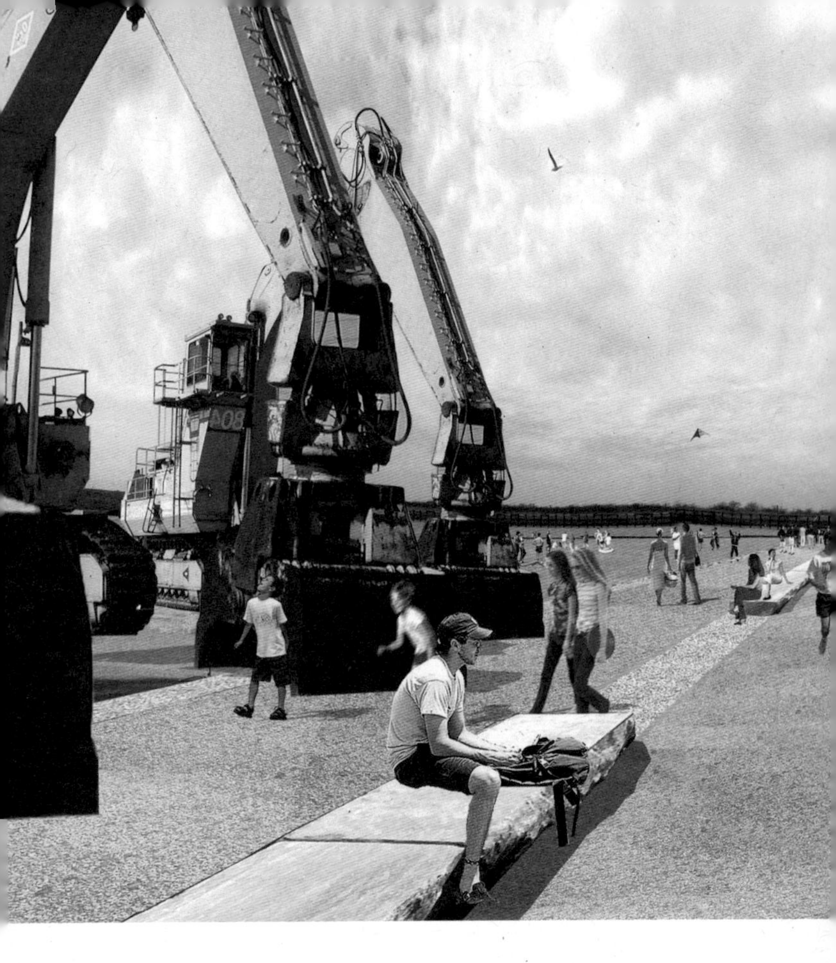

 Lifescape at Fresh Kills, Staten Island: barge garden. Image courtesy of Field Operations.

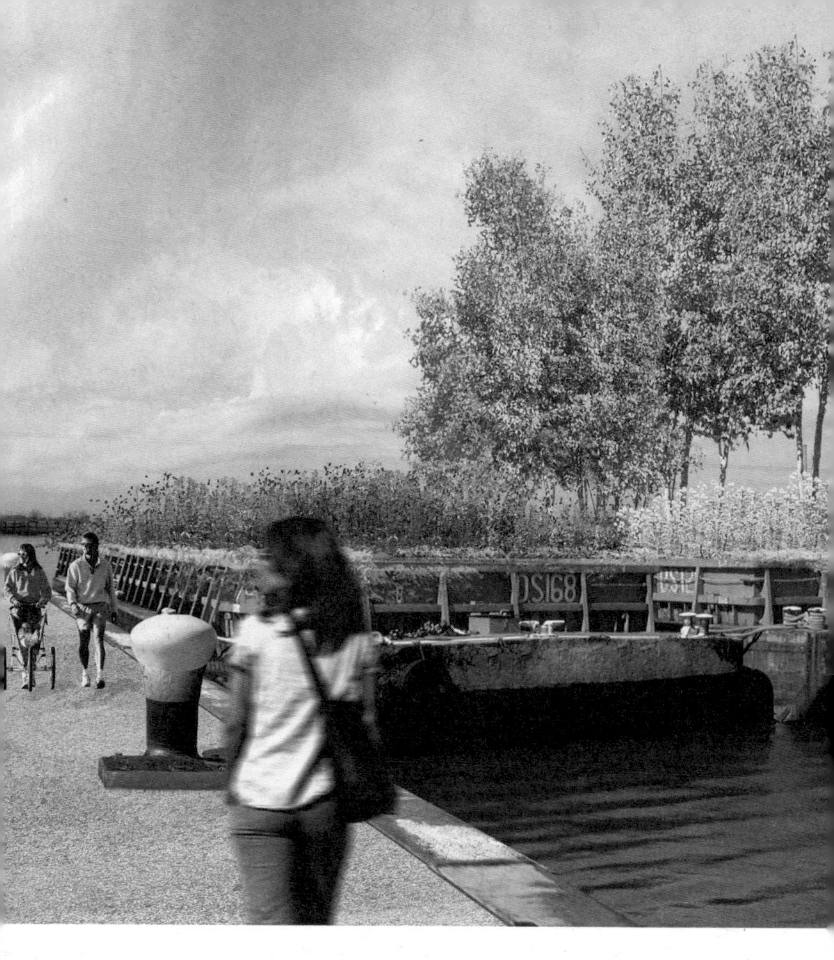
DS168

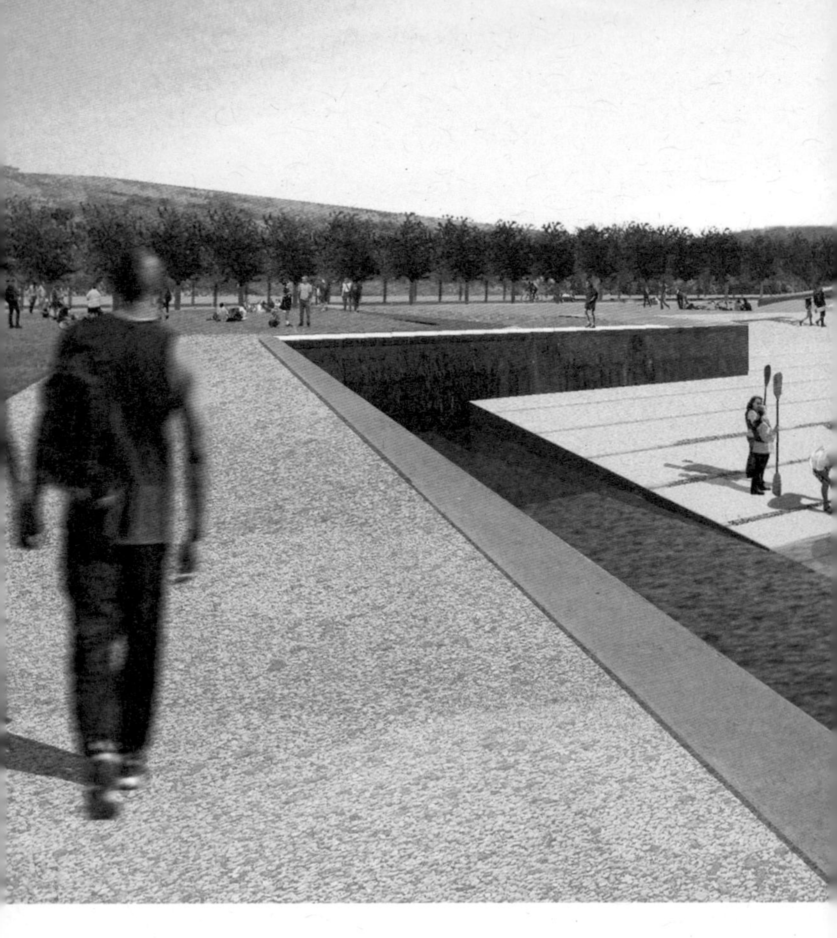

 Lifescape at Fresh Kills, Staten Island: boat launch. Image courtesy of Field Operations.

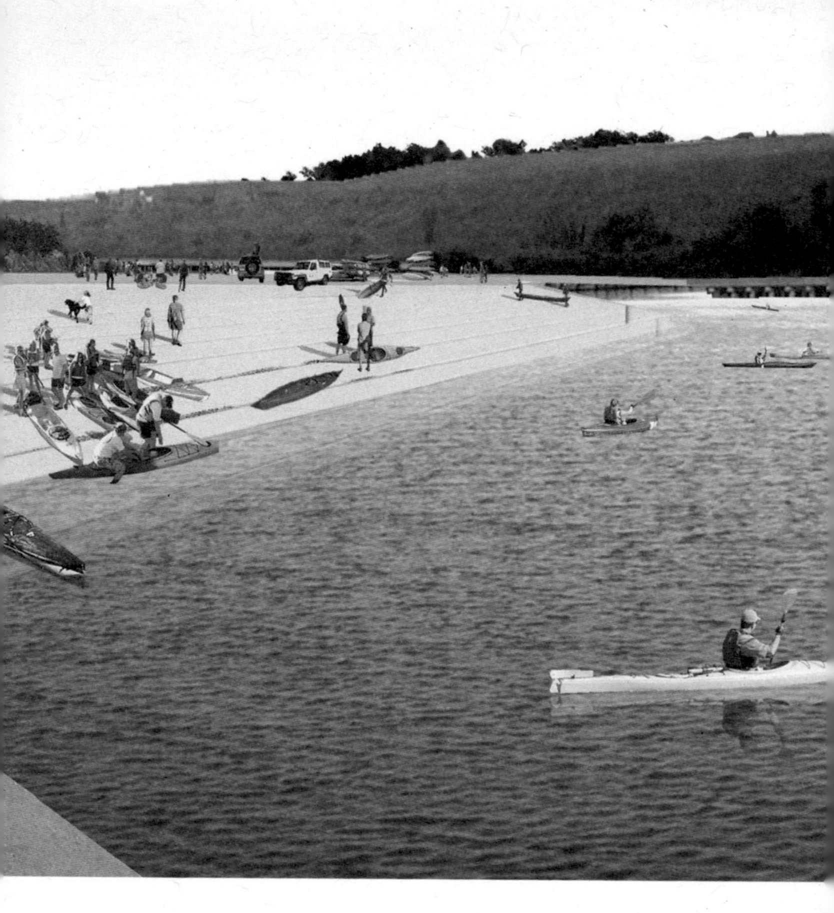

between the mounds. Recreational areas for inland sports like mountain biking and water's-edge activities, such as kayaking and canoeing, will also be included in the short-term plan. Yet these details are really not the point. For Field Operations, the core aim of Lifescape is to seed the evolution of a series of living spaces, products of locally emergent ecologies which develop over time. These spaces will inevitably be used and interpreted in unforeseeable ways by those who are drawn to their unfolding.

On the western coast of India, Mumbai sits in a natural harbor on the Arabian Sea. It is the center of the country's commerce, with thriving banking and entertainment industries. Mumbai's port is the largest on the subcontinent, and handles some 40 percent of India's trade. With a metropolitan population of close to 20 million (and the city proper at about 14 million), Mumbai is the second-largest city in the world, after Tokyo. It is also home to a significant slum population, with more than 7 million residents living in extreme poverty, many of them located along the Mithi River in the settlement of Dharavi, Asia's largest slum. As a result, the Mithi is severely contaminated with both human and household waste. Compounding the public health risk are the seasonal monsoon rains that inundate Mumbai.

Following the catastrophic flood of 2005 in which hundreds perished when the entire season's rain fell in one day (944 mm), the main government strategy for waterfront management was to widen and deepen the Mithi channel, so that it would be able to drain storm water more efficiently. Yet many argue that this strategy is short-sighted. In addition to displacing the poorest residents, it is likely to worsen the flooding without solving the continuing problem of the city's waste, which clogs the drains and outflows. And indeed, the flooding is now worse: given the growing population and the increasing asphalt and concrete surface area, it takes much less water to flood Mumbai today. Coupled with predicted sea level rise, the situation will deteriorate significantly.

To confront the myriad problems with the Mithi River in the context of Mumbai's complex landscape and waterfront, the urban design firm Mathur and da Cunha have departed from the conventional approaches taken by both the current Indian authorities and by the British before them. Their project aims to recalibrate the river from sewer to estuary, and in the process, reconceptualize Mumbai as a resilient, water-adapted and water-dependent city— no longer a seafront port that fears the monsoons, but an aqueous terrain that welcomes and is sustained by the rains and the sea, in synergy.

Since the time of British rule, Mumbai has been fighting both the sea and the monsoon season's winds, heavy rain, and flooding. The British began the tradition of "shoring up," building seawalls and embankments, reclaiming land from the sea while at the same time, in theory, flood-proofing the city with deeper channels, overflow ditches, and concrete levees. All of this was designed to keep the sea at bay and to drain floodwaters from the land as quickly as possible. Current government initiatives do much the same, as Mathur and da Cunha note, "'training' it with walls to conform as much to two lines on a map as to a channel on the ground." But as a city within an estuary, Mumbai is held to the rhythms of the sea: the tidal influence moves upstream twice daily into the Mithi and "exhales" back out again to the sea, carrying with it the detritus of the city. When the ocean swells, or the rains come, the natural process of flushing is exacerbated by severe flooding, and the city quite literally steeps in its own waste. Mathur and da Cunha suggest that the only sustainable course is that Mumbai be returned to its natural tendency to soak, rather than to flood.

According to Mathur and da Cunha, much of Mumbai's problem stems from the conventional perception of the monsoon as "foul weather"—to be feared and prepared for as if it were an enemy attack in a long-waged war. Rather than seeing the city as field for battle—using maps on which the river and its channels are represented as lines and fronts—they propose a fundamental shift from the military bird's-eye view to a ground-level perspective. Viewing the city at eye level from the river's edge, SOAK proposes a series of design "initiations," speculative interventions at various points on the Mithi River, in which earthworks, vegetated berms, and gentle swales could form natural sills and corrugated surfaces that facilitate the gradual spillage, holding, percolation, infiltration, and slow absorption of water, as well as effluent.

In so doing, Mathur and da Cunha reconceive the Mithi from a closed channel with predictable flows to an open field. Each of their twelve proposed interventions works to resolve the problem of flood through transforming the Mithi basin at specific points and as a whole, taking away old lines, territories, and land uses and installing a more complex system, defined by dynamic change, adaptability, and ultimately, in the healthiest state, resilience. Their project is a transition of the mind as much as the land: Mathur and da Cunha's SOAK challenges the colonial control of a landscape delineated by categories of use and instead reaffirms the openness and fluidity of the interstitial space between land and sea, point and line, water and wave. In this space, Mumbai is not a brittle front in the battle between land and water, but a resilient, absorbent city defined by its ecology, alternating between its twin modalities of floating and soaking.

Stoss Landscape Urbanism's RIVER+CITY+LIFE, a design proposal for Toronto's Donlands, is a similarly speculative plan to engage the ecologies of a river within an urban waterfront context. The project was runner-up in a 2007 international design competition sponsored by WATERFRONToronto (the agency charged with the regeneration of the Toronto waterfront) whose goal was to reconceive and revitalize a 40 hectare site of post-industrial waterfront in Toronto at the mouth of the Don River. A key part of the city's multibillion dollar waterfront revitalization plans, the site has significant potential as residential and commercial real estate, yet suffers from a legacy of contamination and a complex pattern of private and public sector land holdings.

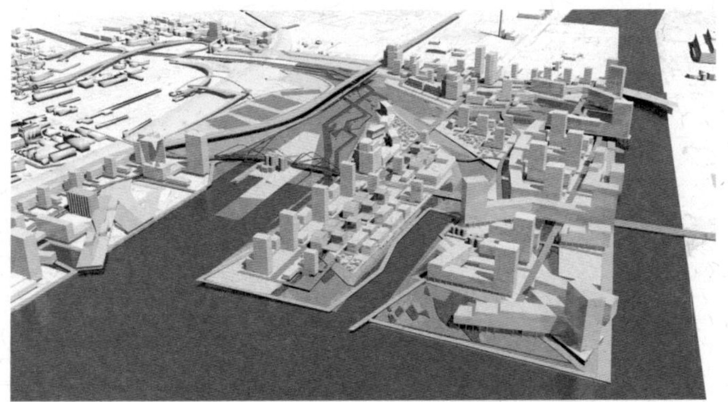

Located in the Great Lakes Basin, on the north shore of Lake Ontario in the industrialized heart of North America, Toronto is a city-region of 5 million and is one of the fastest-growing metropolitan areas in North America. It is also one of the world's most ethno-culturally diverse cities whose social ecology is as complex as the ecologies of its native shoreline and of the Great Lakes–St. Lawrence–Lowland forest that characterizes its landscape. Stoss's design situates the redevelopment at the synergistic intersection of all these ecologies.

The Don River flows from its headwaters in a glacial moraine north of Toronto through the heart of Canada's largest city, bisecting the region and cutting a series of forested ravines before reaching Lake Ontario. The Don River watershed is the largest urbanized watershed in Canada, draining an area of 365 square km. At its outflow, the Don is significantly degraded, having been channelized for the final 2.5 km of its journey to the lake. Stagnant, polluted, and choked with debris, the Don is like many other post-industrial urban waterfront sites: derelict and forgotten, as the armature of the city has all but subsumed it.

In this context, the challenge for design teams was to renaturalize the mouth of the Don River while simultaneously re-engineering the flood plain and creating a new urban edge to the city's downtown.

RIVER+CITY+LIFE, Toronto: plan overview. Image courtesy of Stoss Landscape Urbanism.

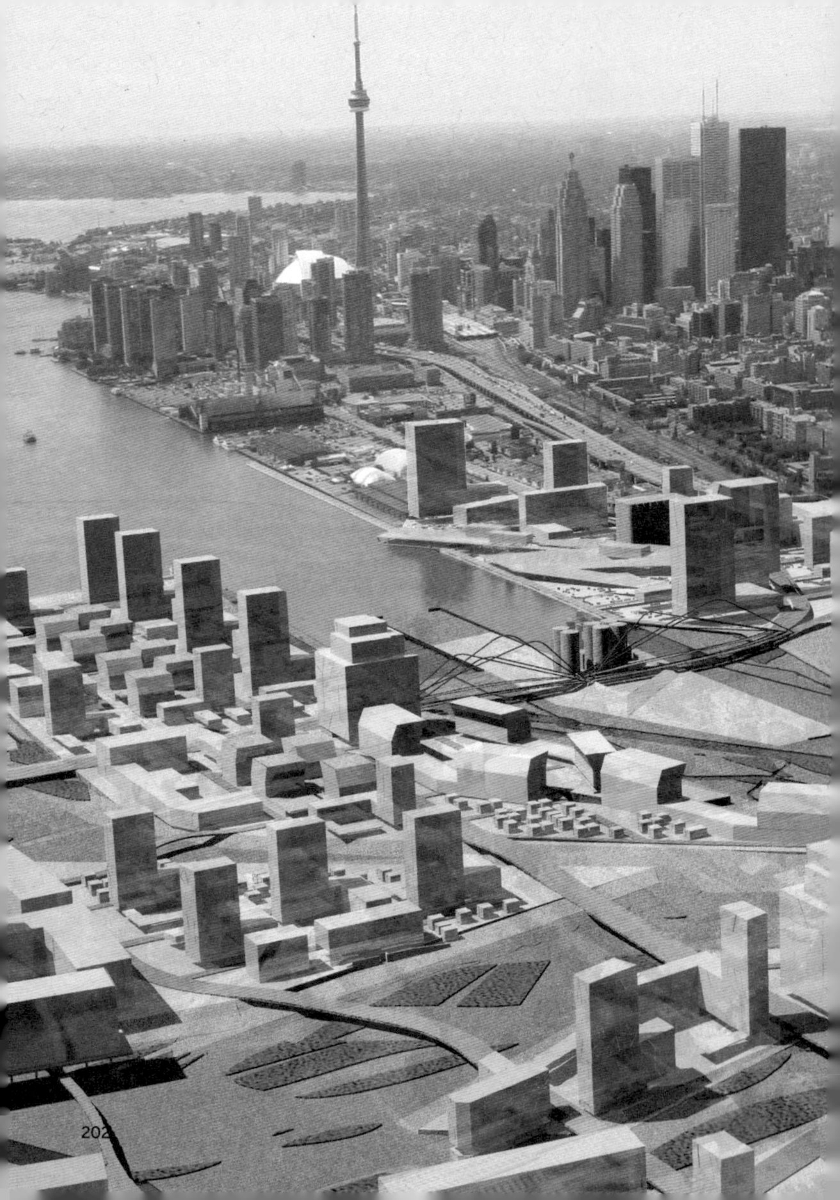

RIVER+CITY+LIFE, Toronto: bird's eye view. Image courtesy of Stoss Landscape Urbanism.

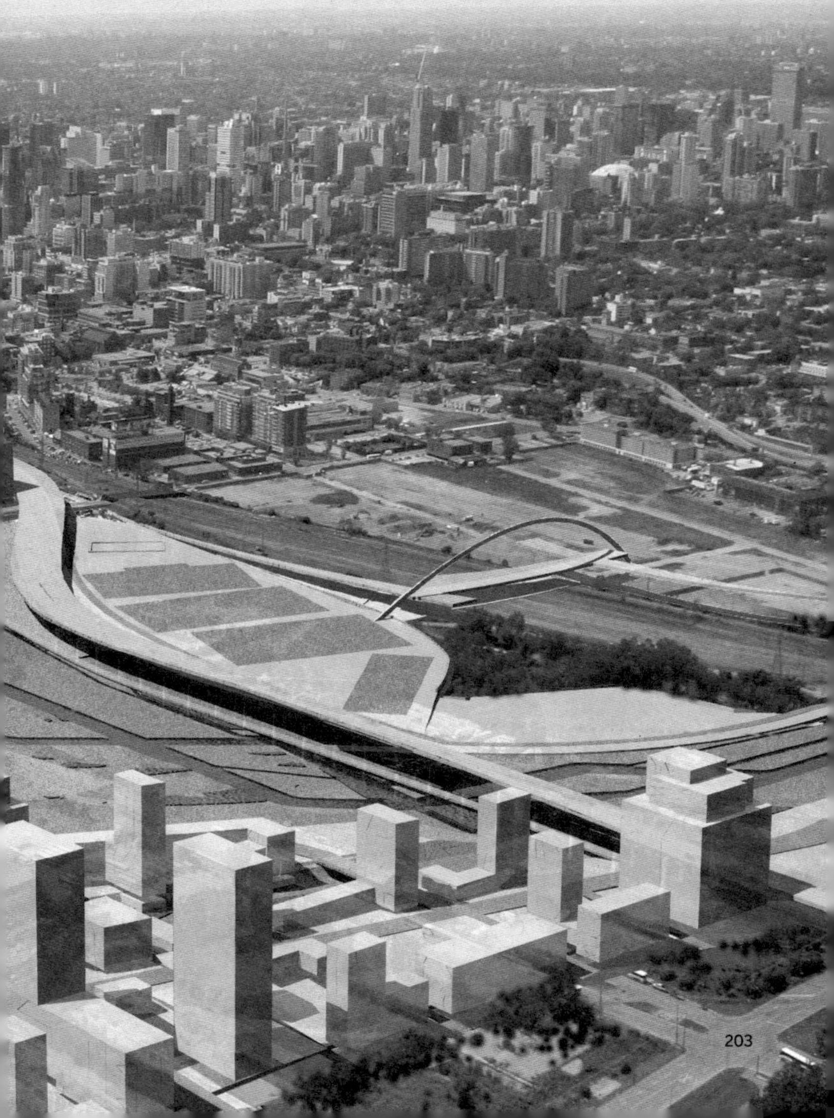

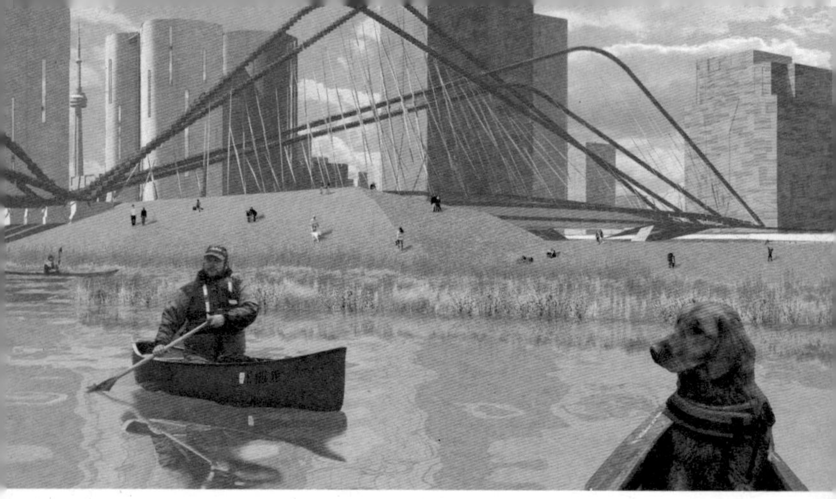

Working at the confluence of the urban core and the underused central waterfront, the team pursued an adaptive design based on the primacy of the river and its dynamics. Of particular significance is the team's explicit emphasis on building resilience, to be achieved by recalibrating the mouth of the river and its floodplain as a new estuary—not a restored estuary, but a landscape transformed through the creation of a new river channel and "river spits"—sculpted landforms, able to withstand changing lake levels and seasonal flooding, that will also provide new spaces for recreation, education, and housing.

By proposing a new set of integrated ecologies for the site, organized principally by the river and its own innate hydrology, the Stoss plan "puts the river first." This represents a complete reversal of the convention of the past one and a half centuries of straightening, channelizing, and deepening the river for the economic benefit of the citizens of Toronto. Centered on renewal rather than restoration, the design strategies involve adaptation to occasional flooding, mediation between native and alien species, and a thick layering of habitats

 RIVER+CITY+LIFE, Toronto: marsh views. Image courtesy of Stoss Landscape Urbanism.

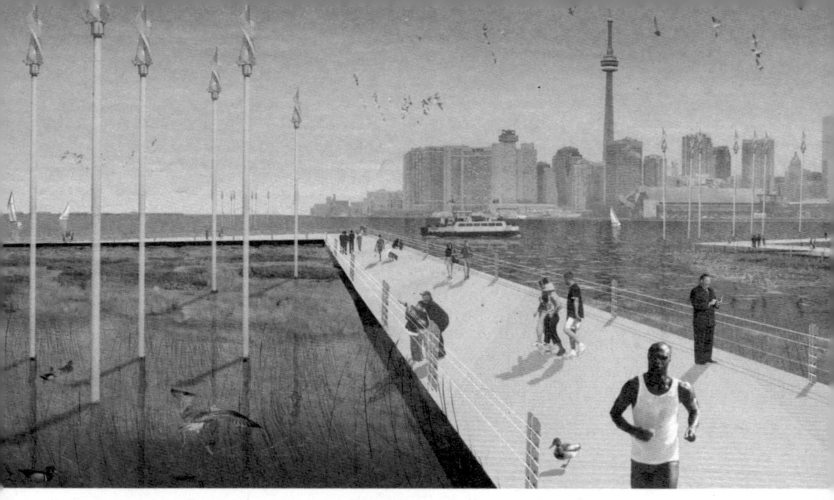

and edges: some cultural, others natural; some seasonal and others permanent. The result is a design that weaves a resilient urban tapestry of public amenity, urban edge, and ecological performance. In doing so, this proposal reconceives the city as a hybrid cultural–natural space, and sets in motion a long-term evolutionary process in which new ecologies are encouraged rather than suppressed.

Lifescape, SOAK, and RIVER+CITY+LIFE each challenge us to think differently about our urban landscapes. These post-industrial projects, in their approaches to the metropolitan waterfront in particular, weave new storylines for design in which the old dualisms of culture and nature are integrated and hybridized, to emerge in dynamic, recombinant expression. Sustainable designs for the contemporary water's edge must necessarily embrace all of the diversity and uncertainty of the confluent forces that continually give it shape. For the millions of people who live, work, and play at the water's edge, these designs offer hope for a renewed, more resilient relationship, in subtle but powerful communication with the waters that sustain us.

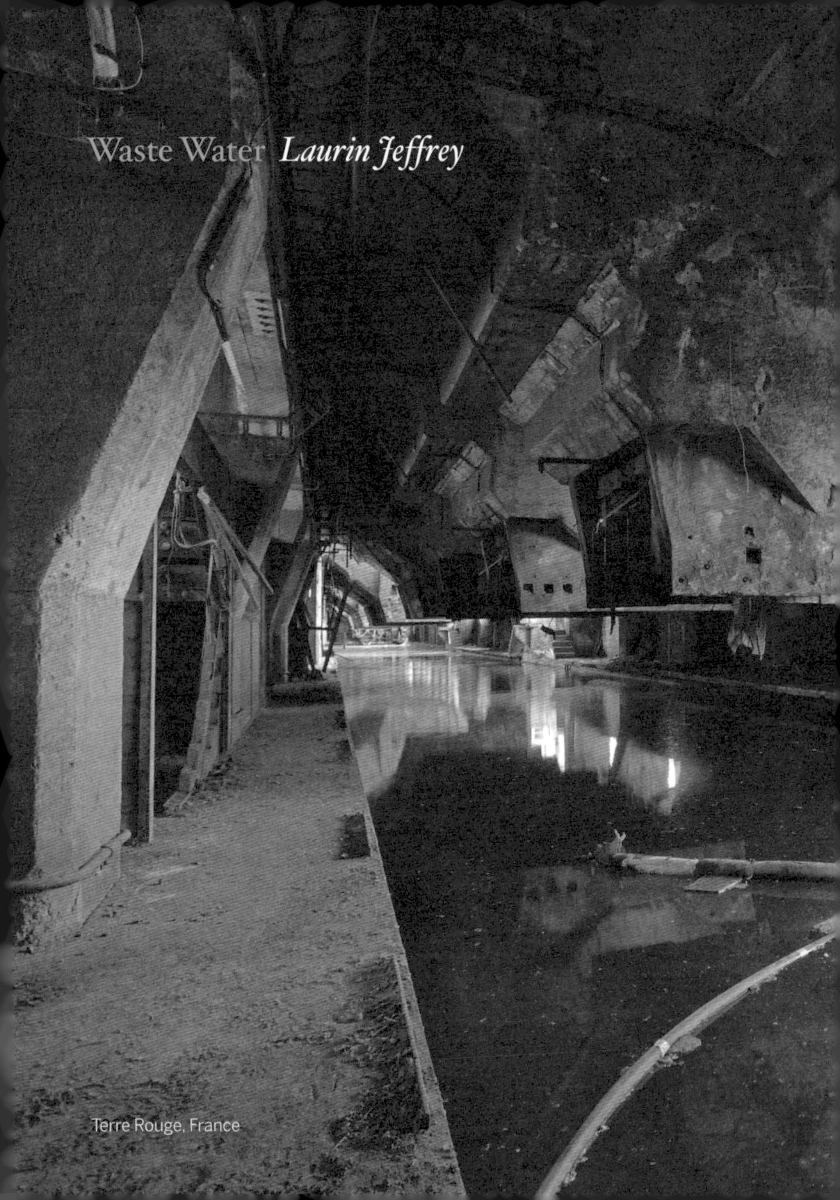

Waste Water *Laurin Jeffrey*

Terre Rouge, France

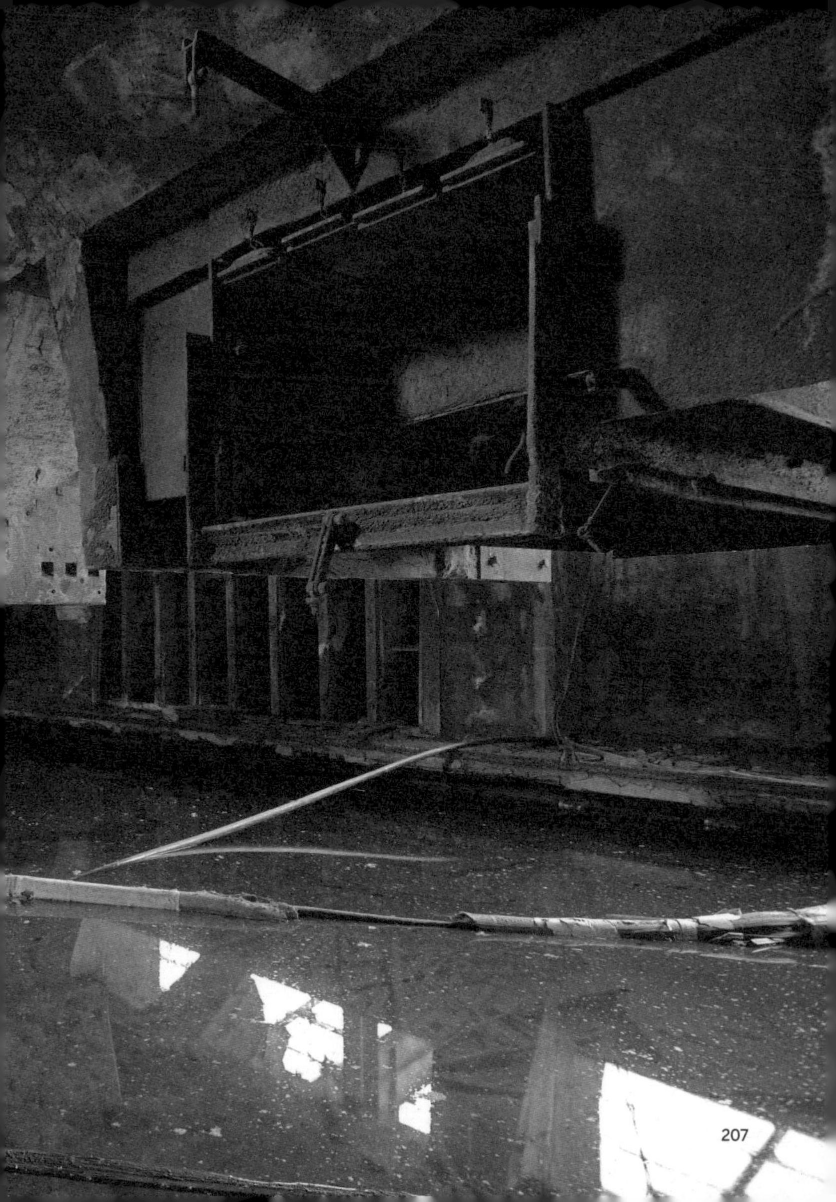

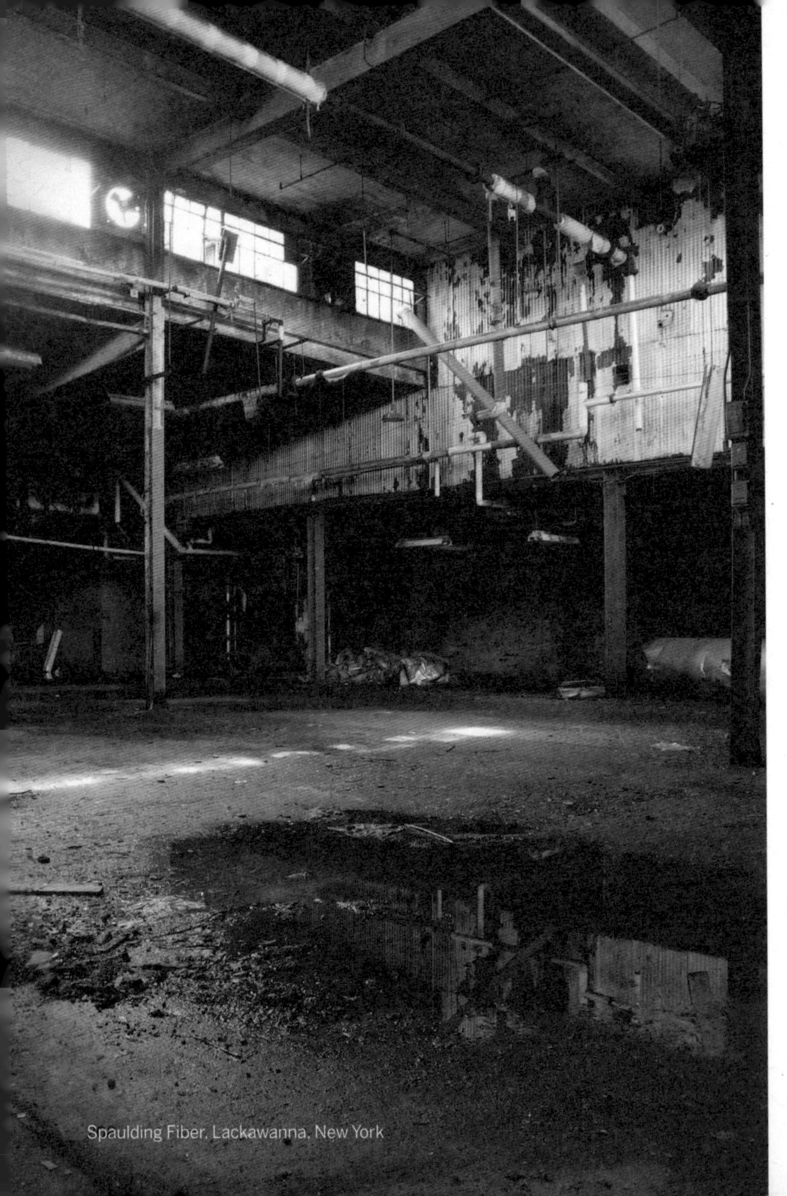

Spaulding Fiber, Lackawanna, New York

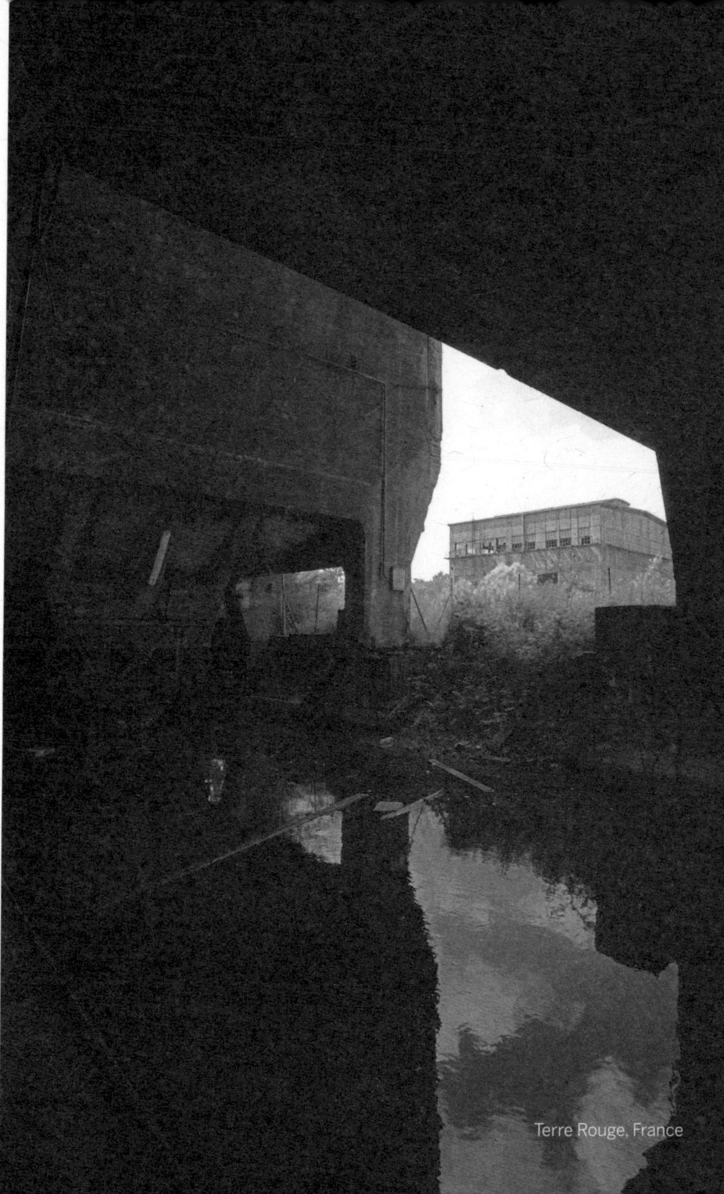
Terre Rouge, France

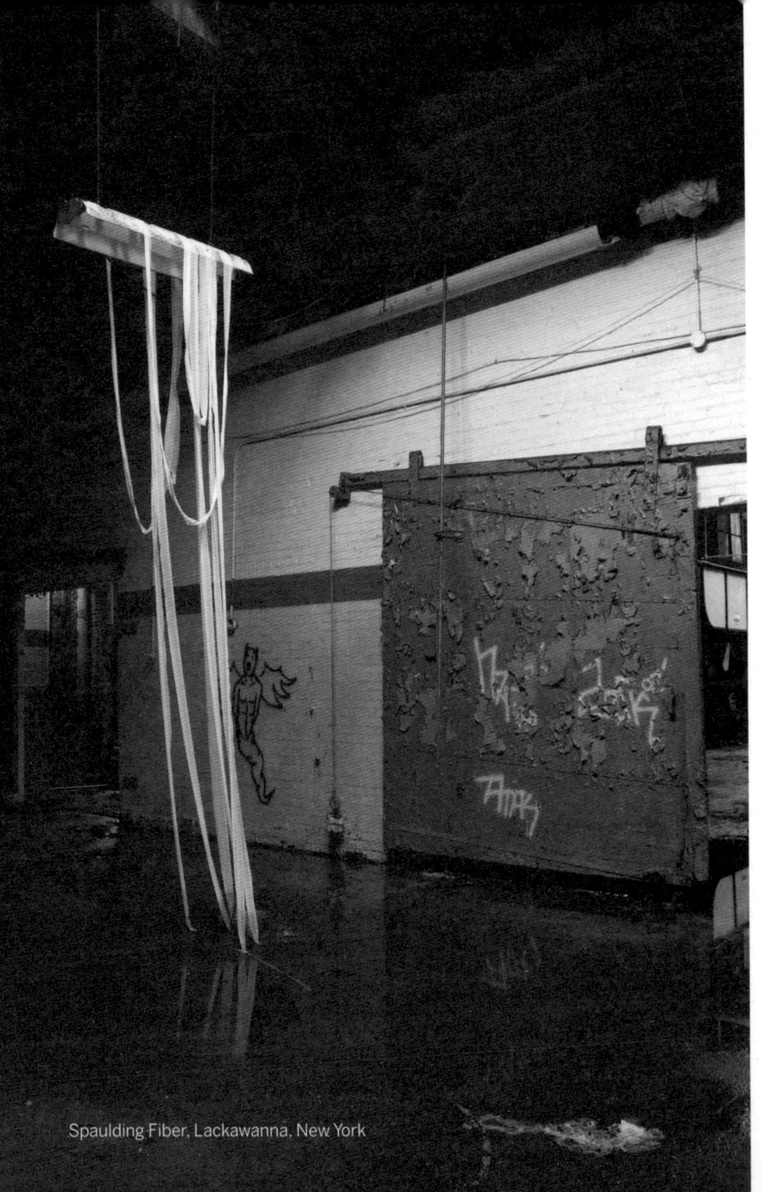

Spaulding Fiber, Lackawanna, New York

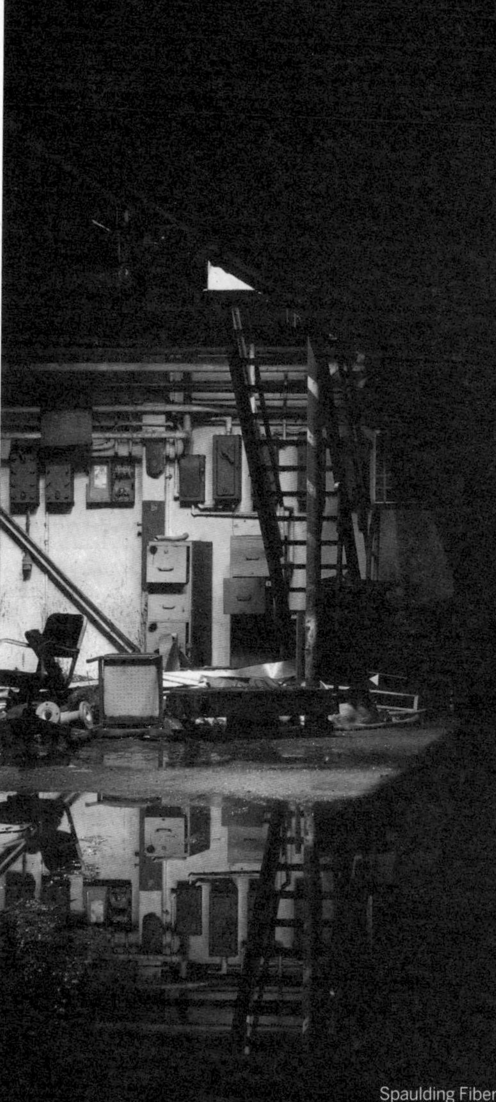

Spaulding Fiber, Lackawanna, New York

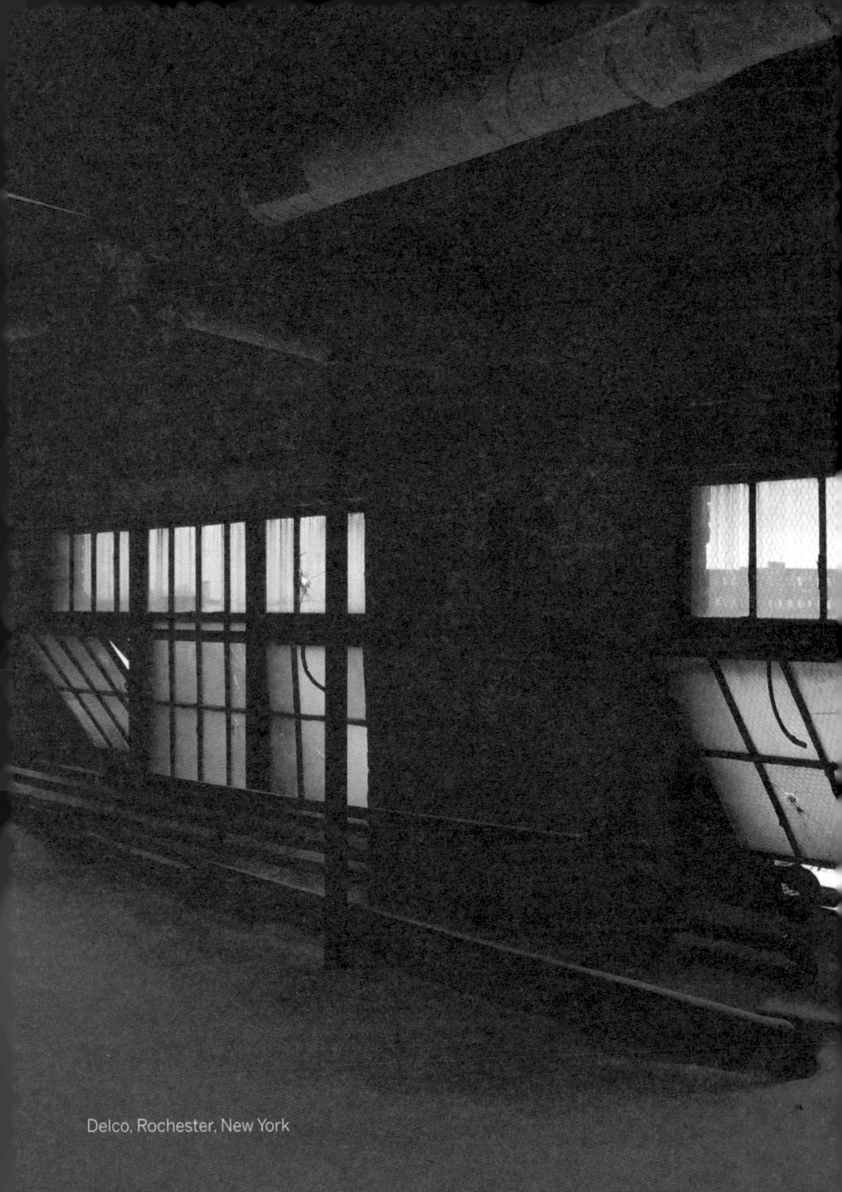

Delco, Rochester, New York

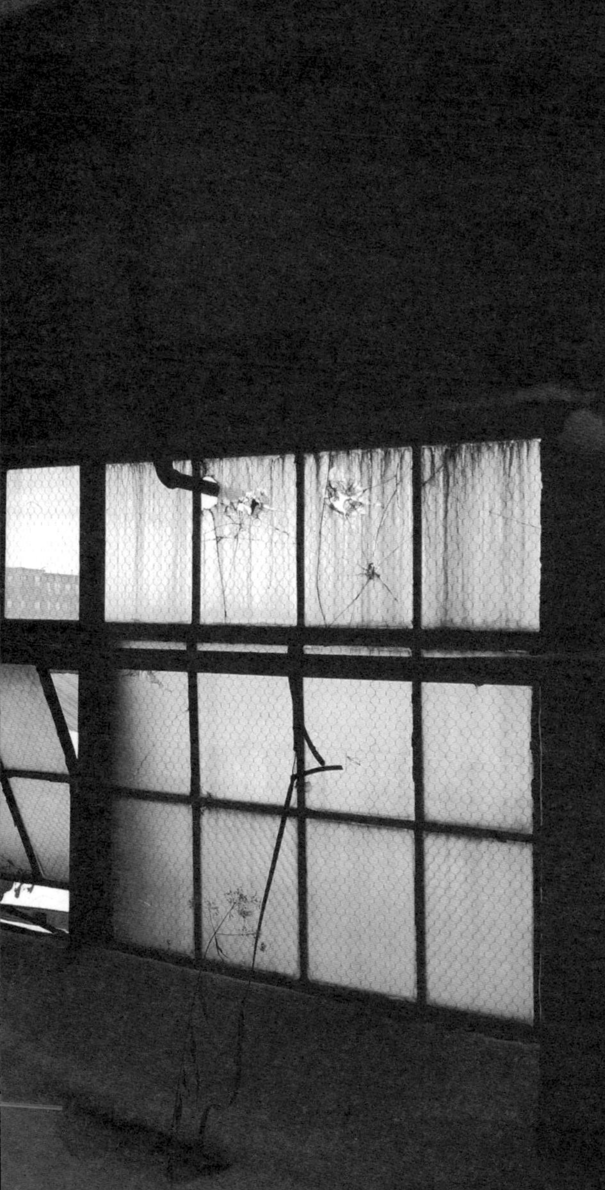

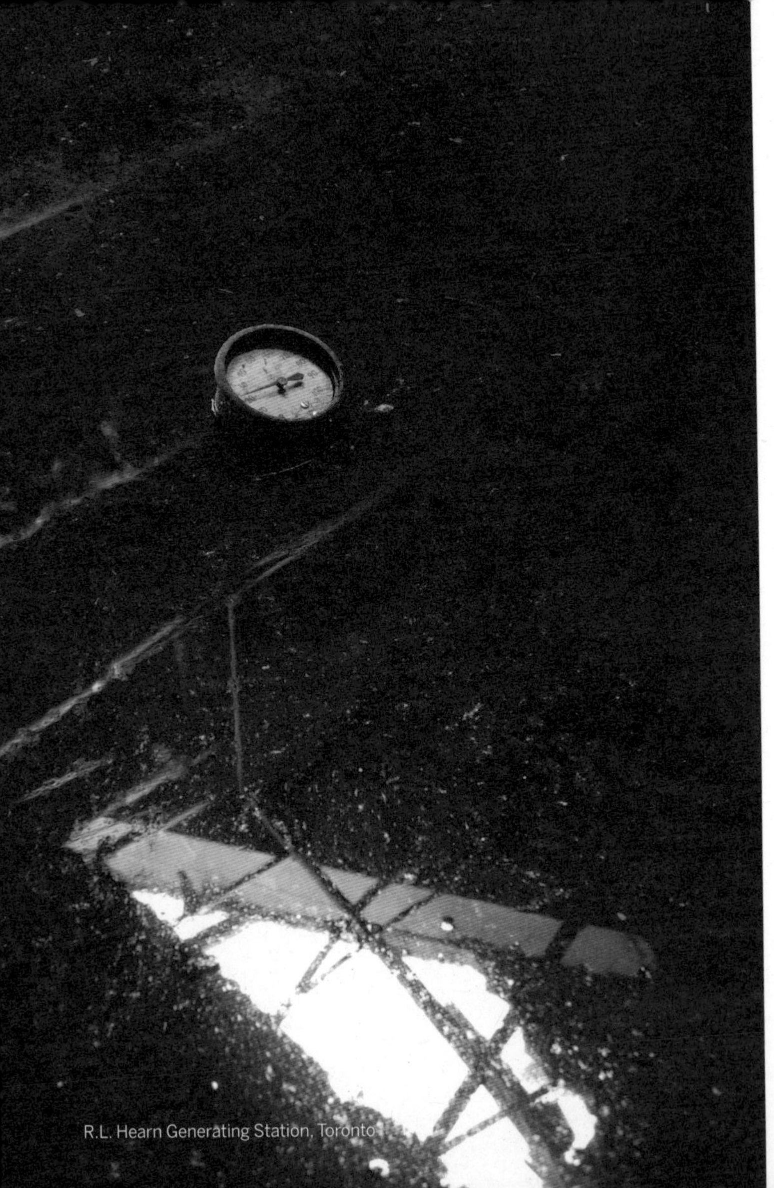

R.L. Hearn Generating Station, Toronto

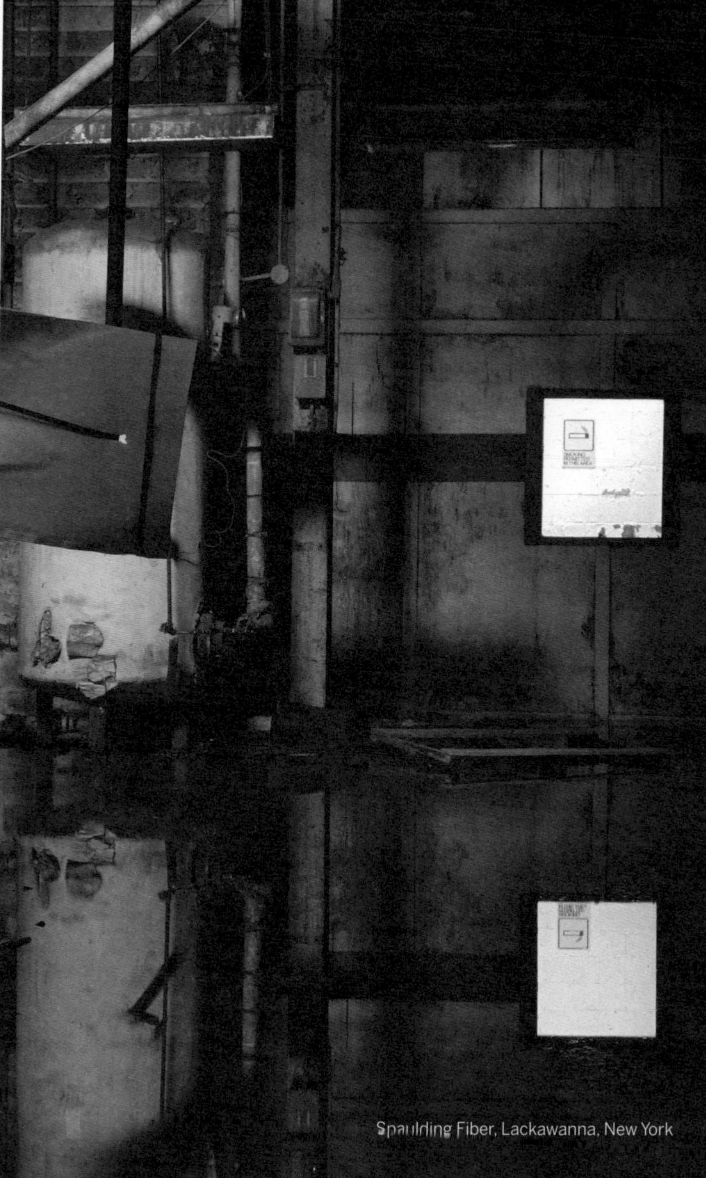

Spaulding Fiber, Lackawanna, New York

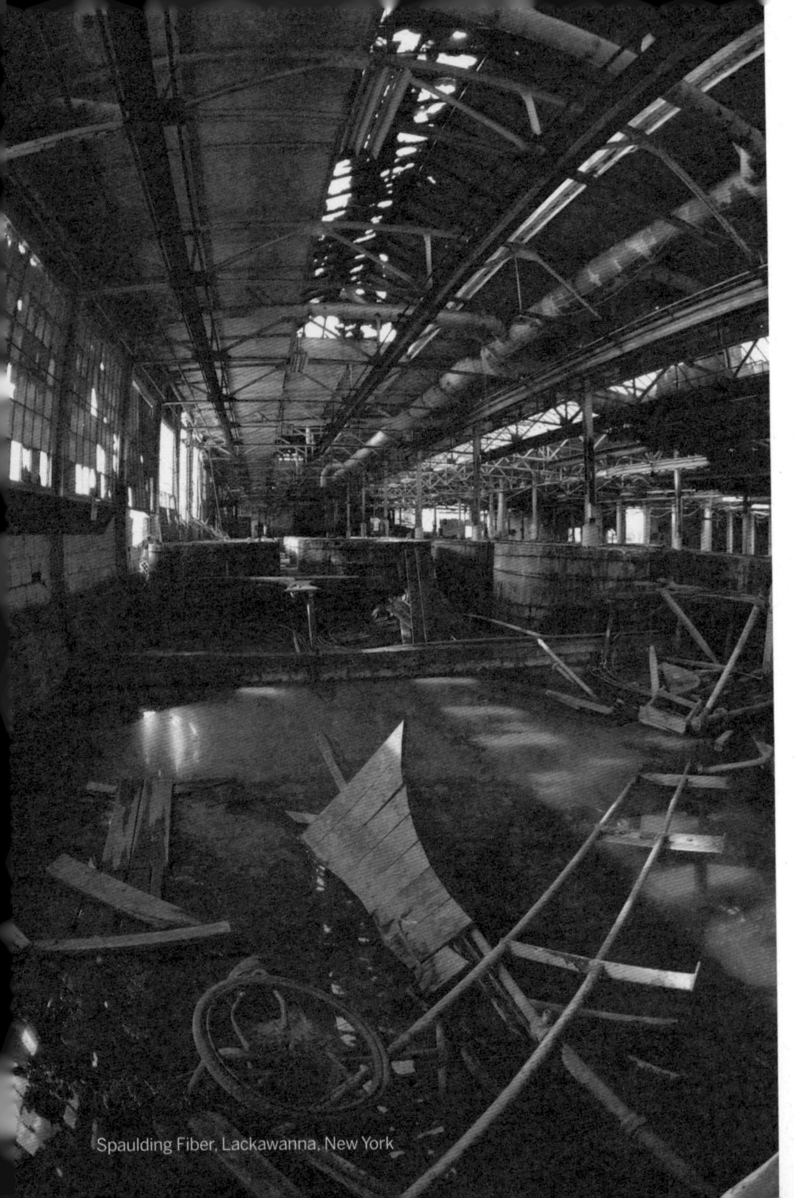
Spaulding Fiber, Lackawanna, New York

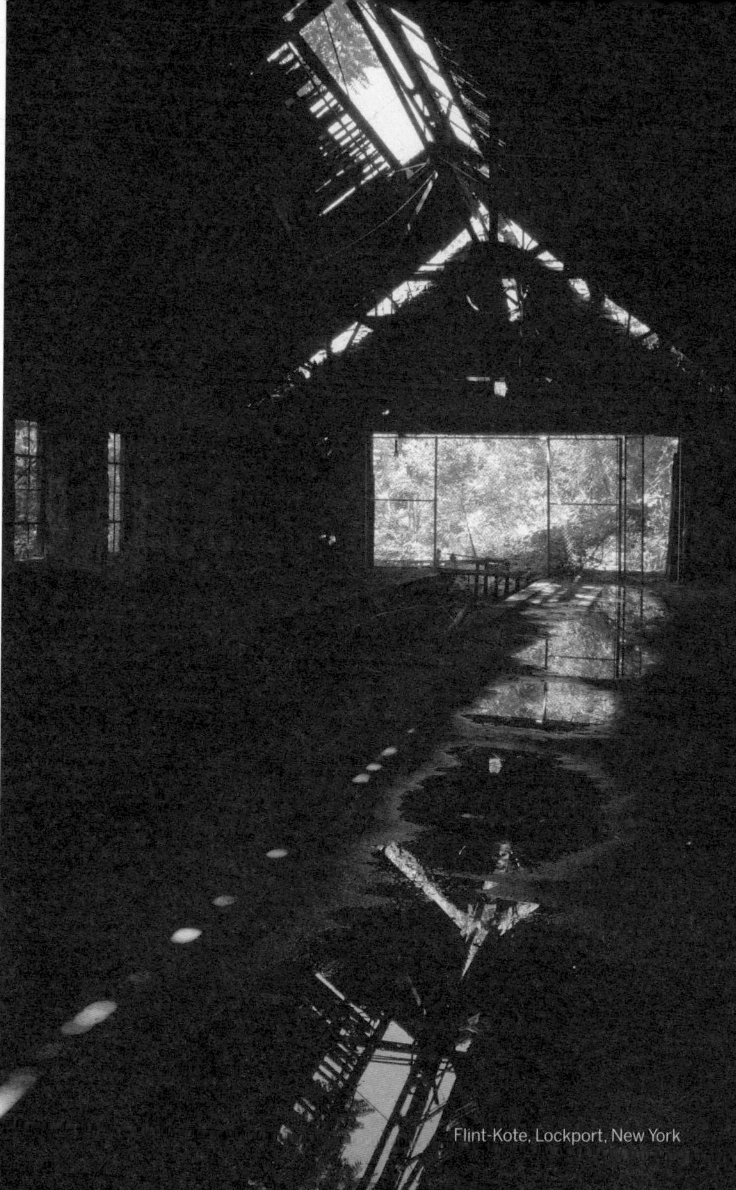

Flint-Kote, Lockport, New York

American Canning, Simcoe, Ontario

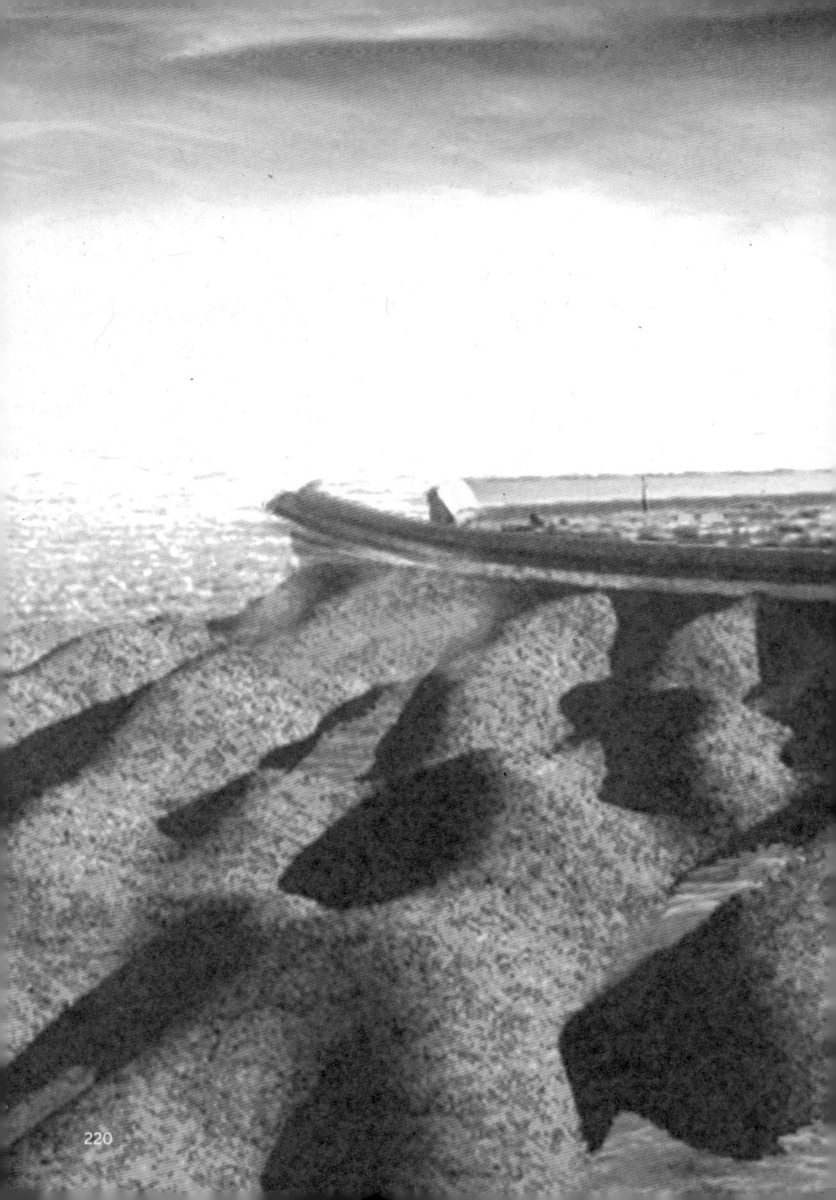

Acque Alte: Transforming the Venetian Lagoon

Colin Ripley, Geoffrey Thün, & Kathy Velikov

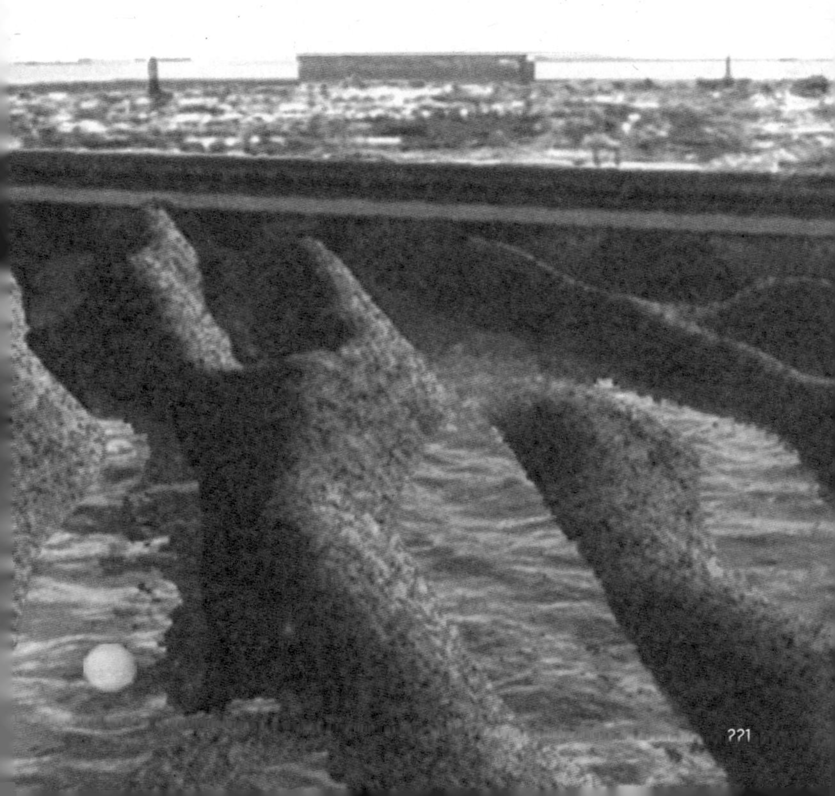

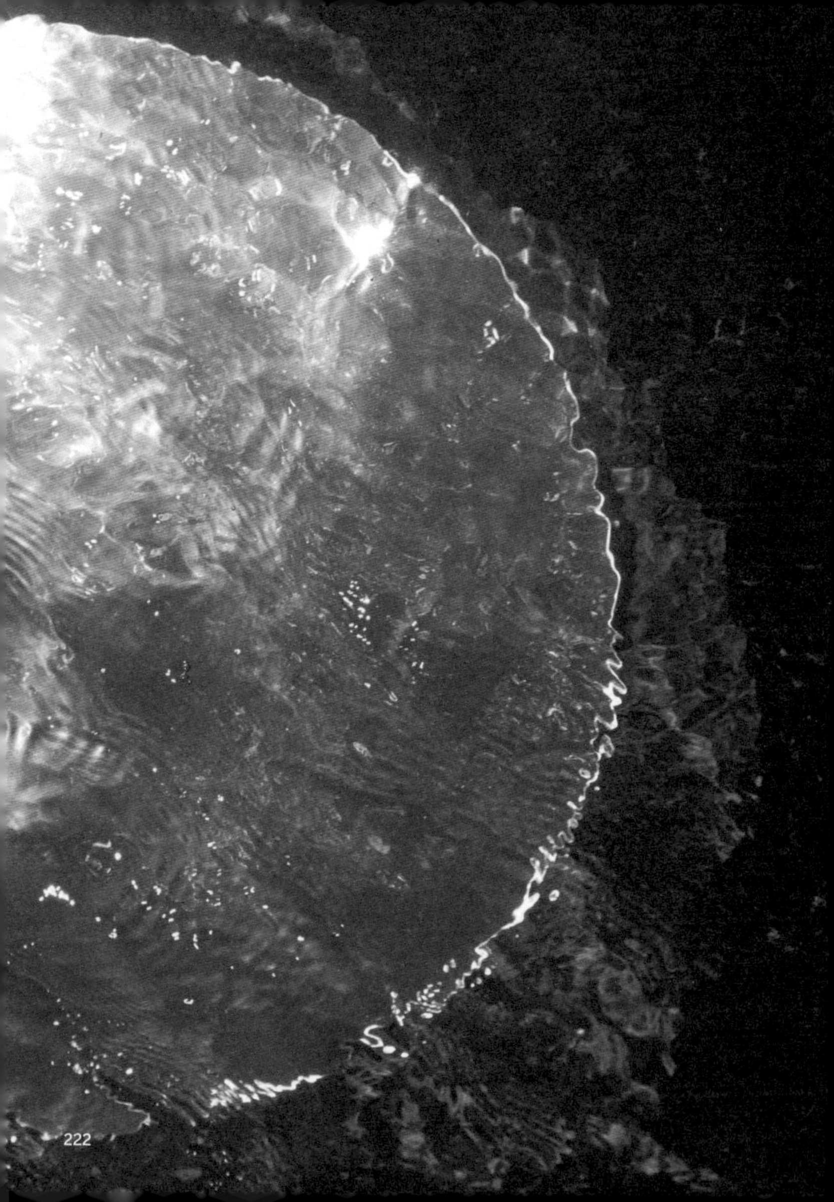

Water's relation to architecture has a long history. The city of Venice marks the quintessential marriage of architecture and water; however, this relationship turns out to be a precarious one—sustained only through an immense technological and political tour de force.

Beginning in the 1400s, in order to preserve shipping access to the Adriatic Sea, the rulers of the Venetian Republic decreed military and defensive operations to halt the natural evolution of the Venetian lagoon into coastal marshland. By 1480 the River Brenta had been diverted and no longer emptied its fresh water and silt into the lagoon; by 1683 the River Sile had also been turned aside. As a result, the lagoon's water grew increasingly saline and the makeup of marine life changed. During the same period, the original eight sea inlets along the lagoon's shore were reduced to three and deepened, making the lagoon increasingly subject to sea wave action and erosion. The lagoon has enlarged since 1400; several islands have sunk and the woodlands at their edges have become intertidal flats. In the twentieth century, this erosion was counteracted by intensive land reclamation. Indeed, only through constant hydraulic intervention has the lagoon been prevented from either silting over entirely and joining the Venetian plain, or having its sandbanks and coastline claimed by the Adriatic Sea.

The famous Venice Lagoon is therefore far from a natural phenomenon. Rather, it is a major piece of human-engineered infrastructure, carefully constructed and maintained for the benefit of the city's military and mercantile needs. Unfortunately, this infrastructure is now in a precarious state.

Depleted underground aquifers, compacted silts, and the gradual rise in sea levels have caused the waters of the lagoon to rise almost

22 centimeters relative to the islands between 1908 and 1980, placing the city under ongoing threat of tidal flooding. St Mark's Piazza sits only a few centimeters above the current spring high-tide sea level. The centuries-long struggle to maintain the lagoon continues, with massive infrastructure projects such as MOSE (Modulo Sperimentale Elettromeccanico or Electromechanical Experimental Module), a three-and-a-half-billion-euro floodgate designed to protect the lagoon from a rise in water levels of up to one meter.

But worst-case global climate change predictions foresee a rise in sea level of as much as 14 meters in the next century, a result that would submerge the entire Venetian plateau in the Adriatic like a latter-day Atlantis, easily swamping even MOSE. While the image of a Venice sunk beneath the waves is painful to contemplate, it may be an inevitable result; the ongoing war against the sea may be one that the Venetians cannot win. In this context, a more radical approach is required, to start considering what ecological and economic possibilities might emerge from a flooded lagoon. The lagoon has the potential to serve as a future space for aquacology, with the surface of the water supporting industrial-scale food and energy production. What's more, both the water's surface and the cultural treasures buried beneath it present wonderful opportunities for an expanded tourist industry: the Venice Lagoon Park.

We imagine a future in which Venice proper, its nearby islands of Murano, San Michele, and Giudecca, and a few other culturally significant lagoon islands (as well as Marco Polo airport) will be enclosed by protective ring walls, inside which current water levels will be maintained. Except for these few preserved islands, the Venice Lagoon Park will be primarily an aqueous experience. Constructed using dam-building methods, with locks for boat access, the ring walls will act as a prophylactic barrier. Far from being merely protective devices for the city of Venice, the rings will also form a critical link between the water-based food and energy production and land-based end users, containing facilities for receiving, processing, warehousing,

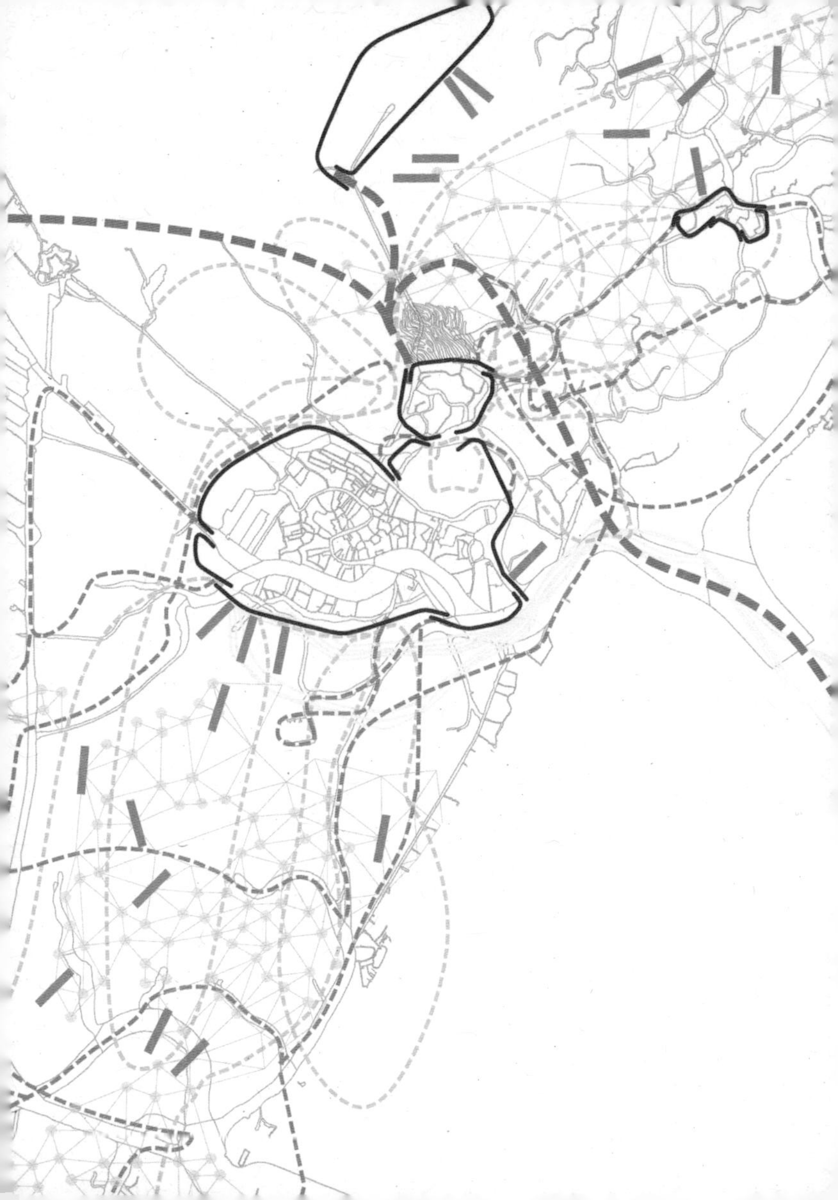

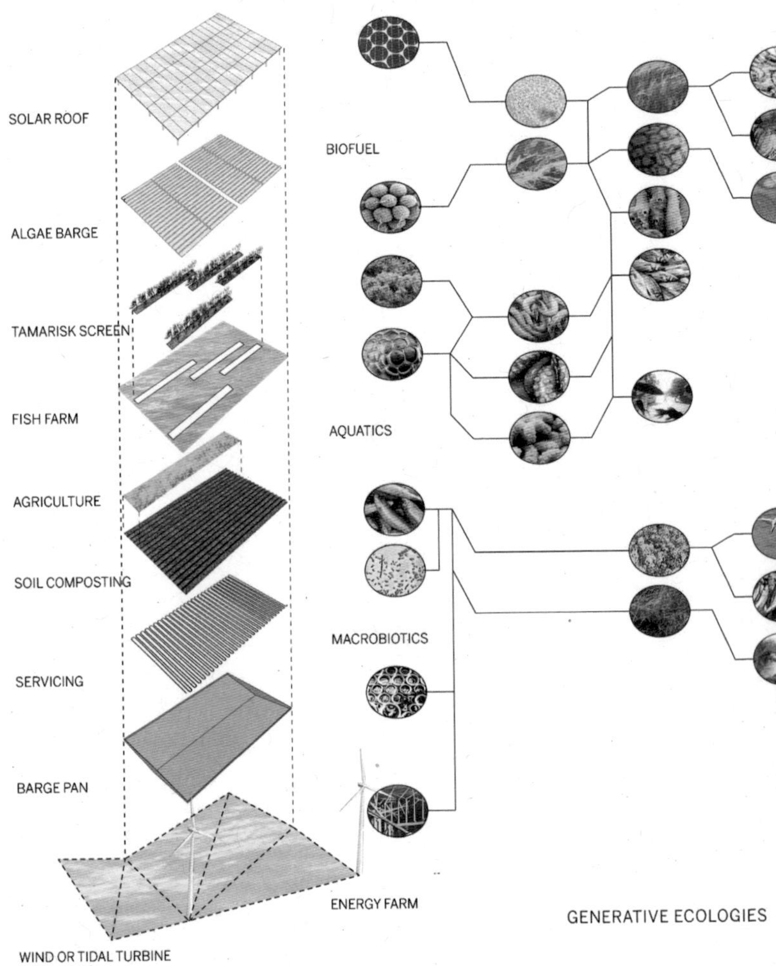

 Venice Lagoon Agricology Components Diagram

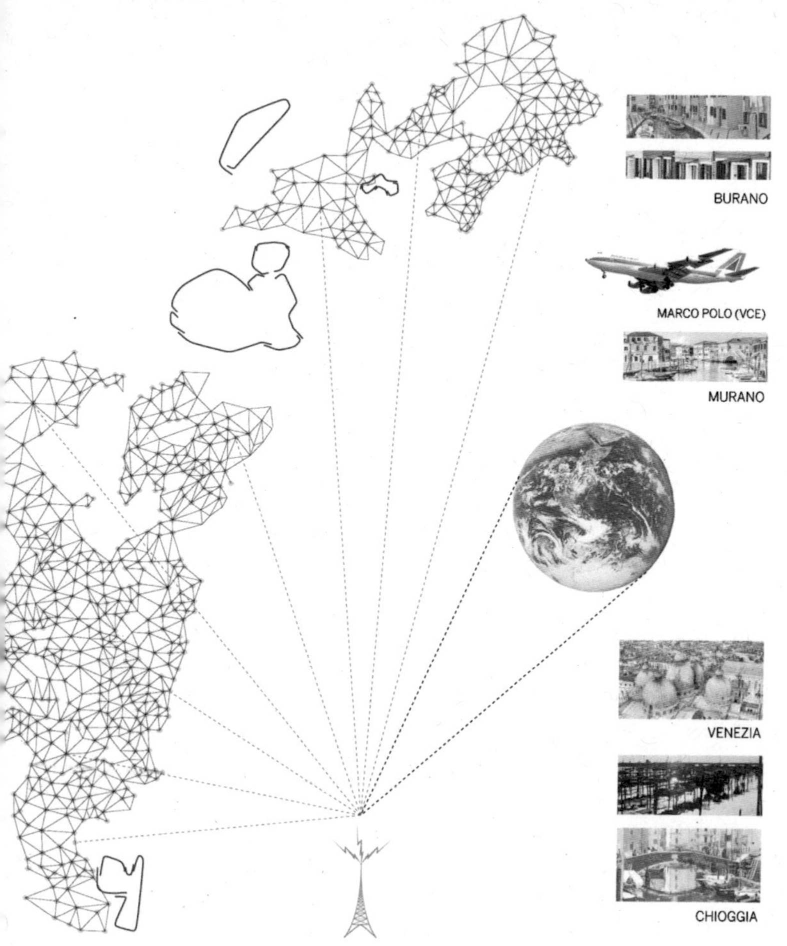

DISTRIBUTED INFOMATICS

GEOMATIC AND INFOMATIC GRID

CULTURAL / CIVIC ASSETS

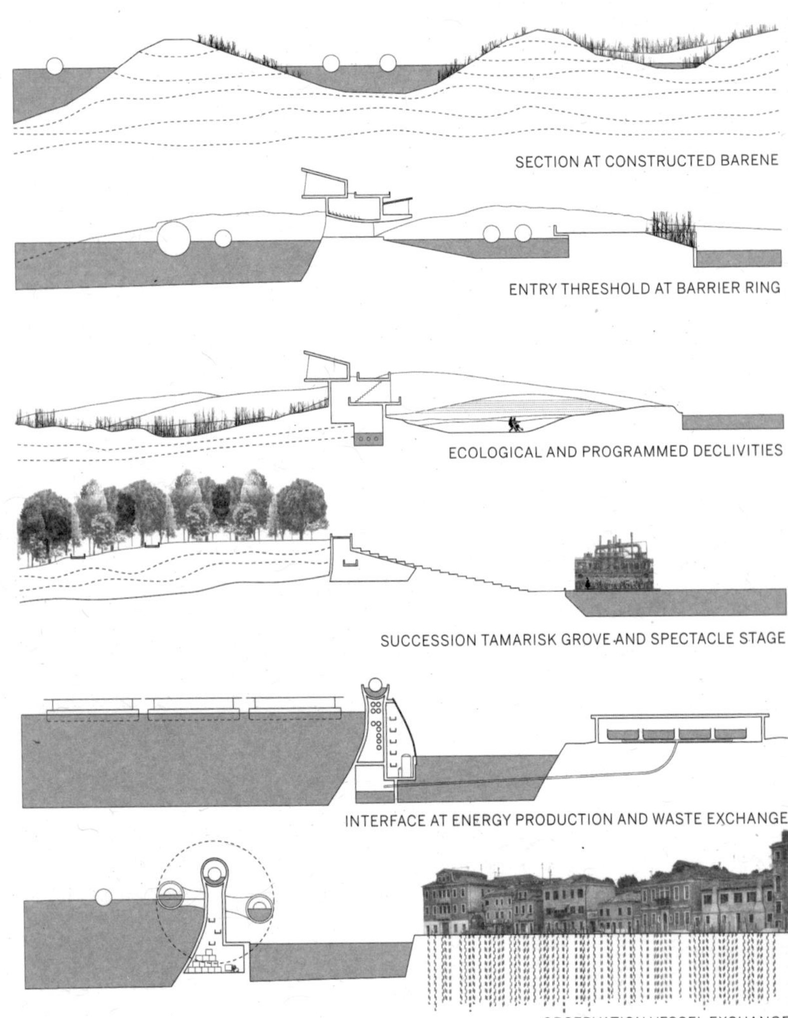

 Prophylactic Barrier Rings: Sections at Ecological Preserve

and shipping the products of the new lagoon agricology. They will become the landing stages for all vessels plying the new cultural and ecological lagoon park, and the gateway to the remaining islands; travelers disembarking on these outer rings will find information centers and sites for renting tourist vessels.

Although the end of daily tidal flushing of the lagoon sandbanks, or *barenes*, will precipitate the death of a number of currently existing ecologies within the lagoon, new ecologies will emerge. Micro-algae, naturally occurring in most expanses of water, will be intensively farmed on a matrix of barges, and used as a source of food and minerals, and as a basis for hydrogen production. Pink micro-algae such as *Haematococcus pluvialis*, which is found naturally in the Adriatic Sea, use the energy of the sun through photosynthesis to fix CO_2, the most significant greenhouse gas, into carbohydrates, which can then be used as a food source. A byproduct of *H. pluvialis* is astaxanthin, a very expensive pigmentation source used in the food and cosmetic industries. The production of astaxanthin will add another niche industry to the Venice Lagoon. Other strains of algae can be used for biofuel production, with yields a hundred times those of corn ethanol production. Still other strains may be used to make hydrogen through photosynthetic water splitting. The result will be a system that can simultaneously grow food, produce clean energy, and reduce atmospheric greenhouse gases.

In addition to this algal-energy industry, other barges will support more traditional hydroponic agriculture and fish hatcheries, as well as offshore wind turbines and solar collection cells. Even MOSE will be retooled to collect tidal energy.

A new landscape in the form of ridges, valleys, and inlets will be created with the material gained from the excavations required to construct the barrier rings. This ecological preserve and study area will be connected to the existing island of Sacca San Mattia (itself an entirely constructed terrain, built up in the middle of the twentieth century, partially with debris from the Venetian glass industry), just

Atop Prophylactic Barrier Ring

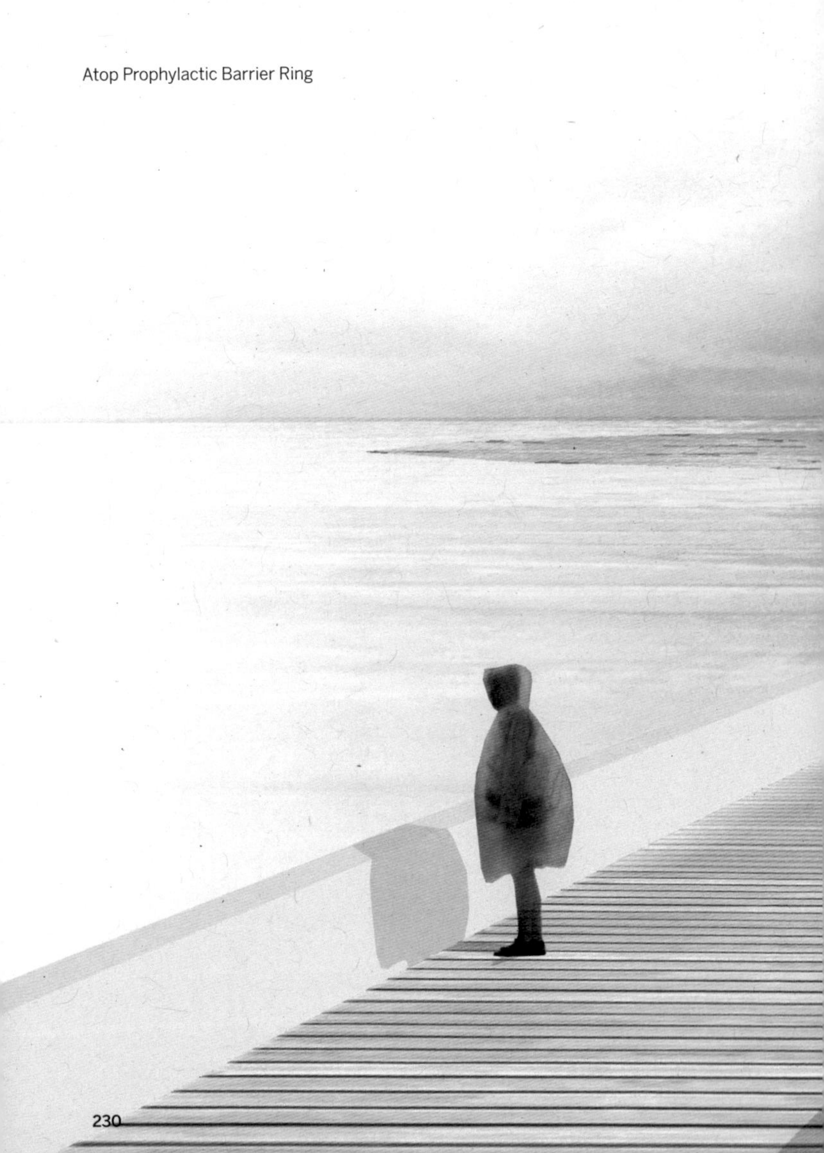

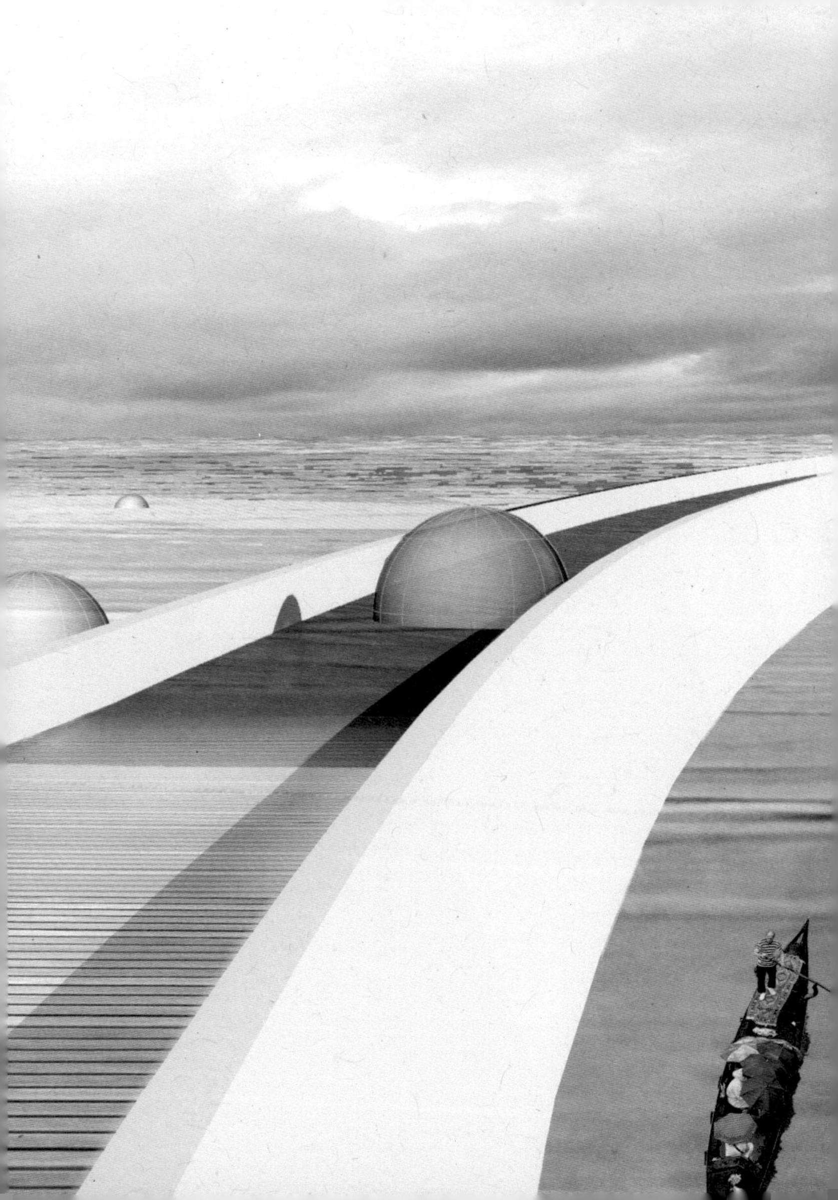

Within Prophylactic Barrier Ring

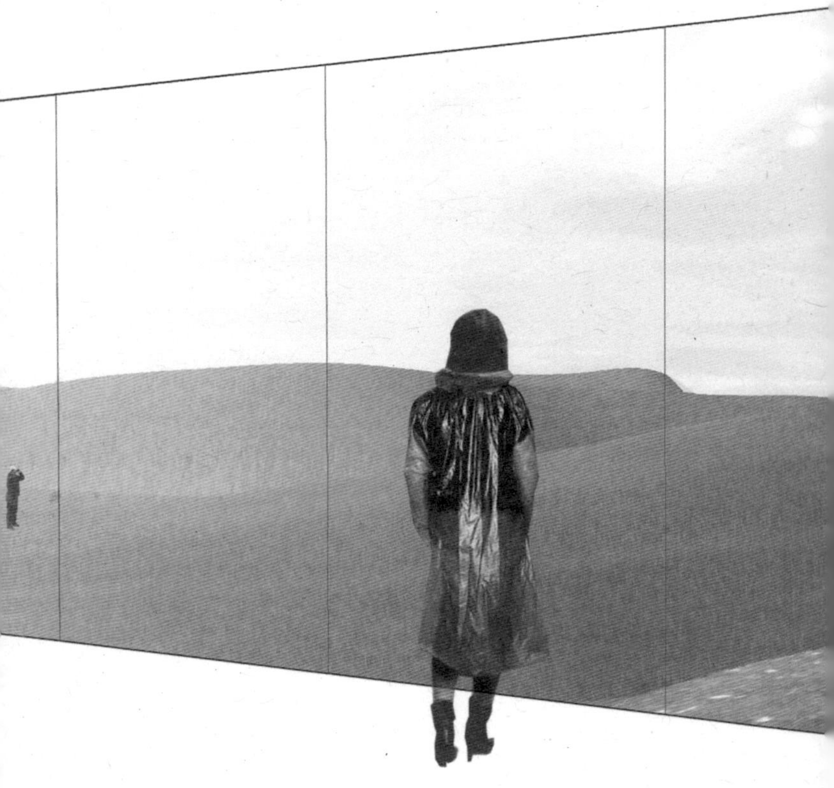

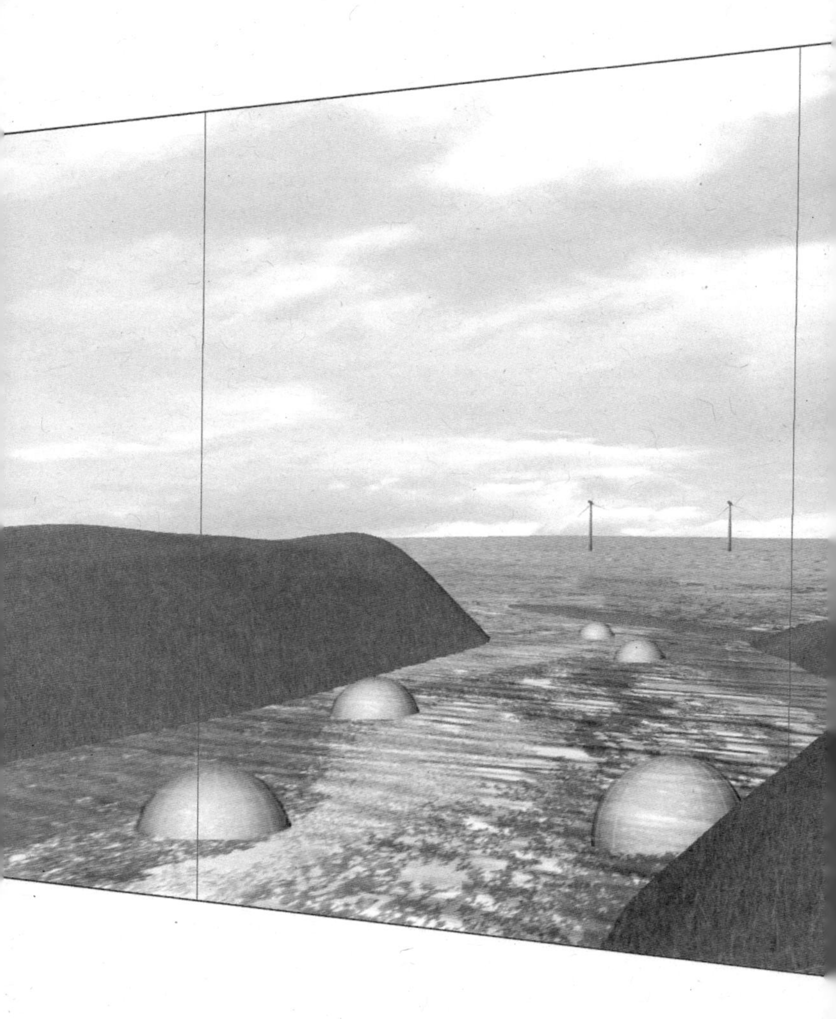

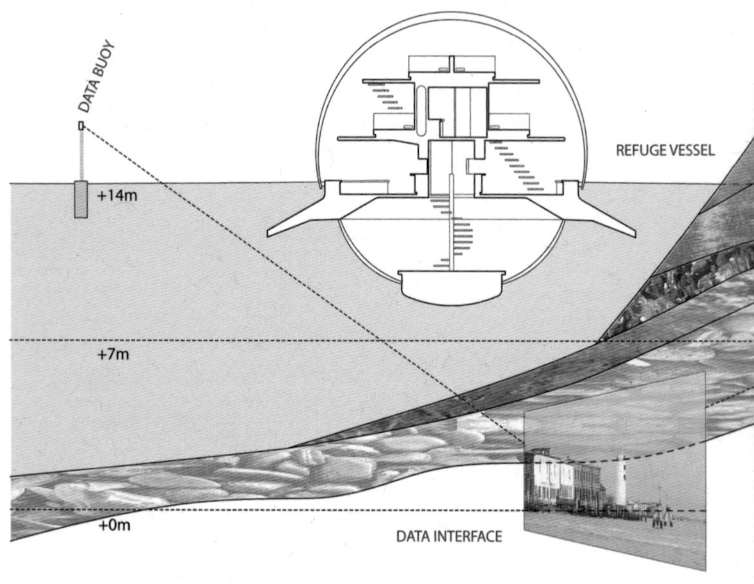

outside its protective ring. The preserve will evolve over time, altered daily by the tides. The southern portion of the new San Mattia, inside the barrier ring, will contain sports fields, recreational circuits, and open-air theatres. Tidal flats and marshlands will be manufactured as a dynamic ecological zone in the northern portion of the island for the preservation and study of marine species.

Taken together, these new infrastructure elements (a submerged city, protected cultural zones, aquacological food and energy industries, ecological preserves) will continue to drive Venice's tourist industry. New facilities will be created to allow visitors to observe and study the historical elements of the Venice Lagoon Park and its floating industrial present. Tourists will board transparent floating bubbles

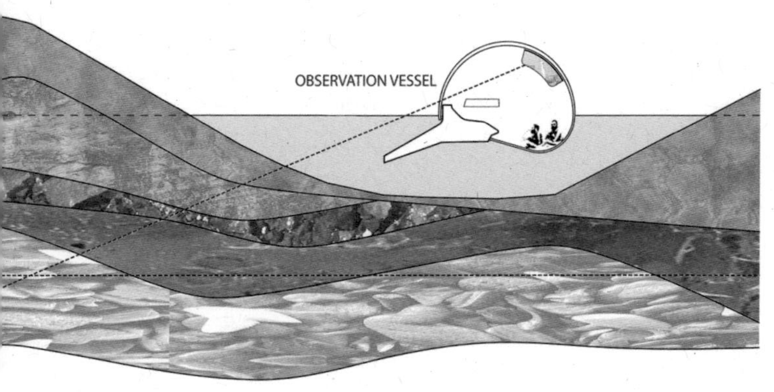

for one or two people in order to view the ecological zone and the submerged portions of the park. In addition to allowing unrestricted views both above and below the water line, these vessels' highly performative skin will provide information, wayfinding, and air filtration. The skin will also act as a hydrogen-producing membrane, using artificial photosynthesis to split water molecules, thereby making the vessels energy self-sufficient.

Larger hotel-like vessels for up to eight people, tethered at nodal points of the geomatic and informatic grid laid over the lagoon, will accommodate longer stays in the Lagoon Park. Artificial photosynthesis will be deployed in the skin of the hotel vessels to provide not only hydrogen but also food, in the form of carbohydrates derived

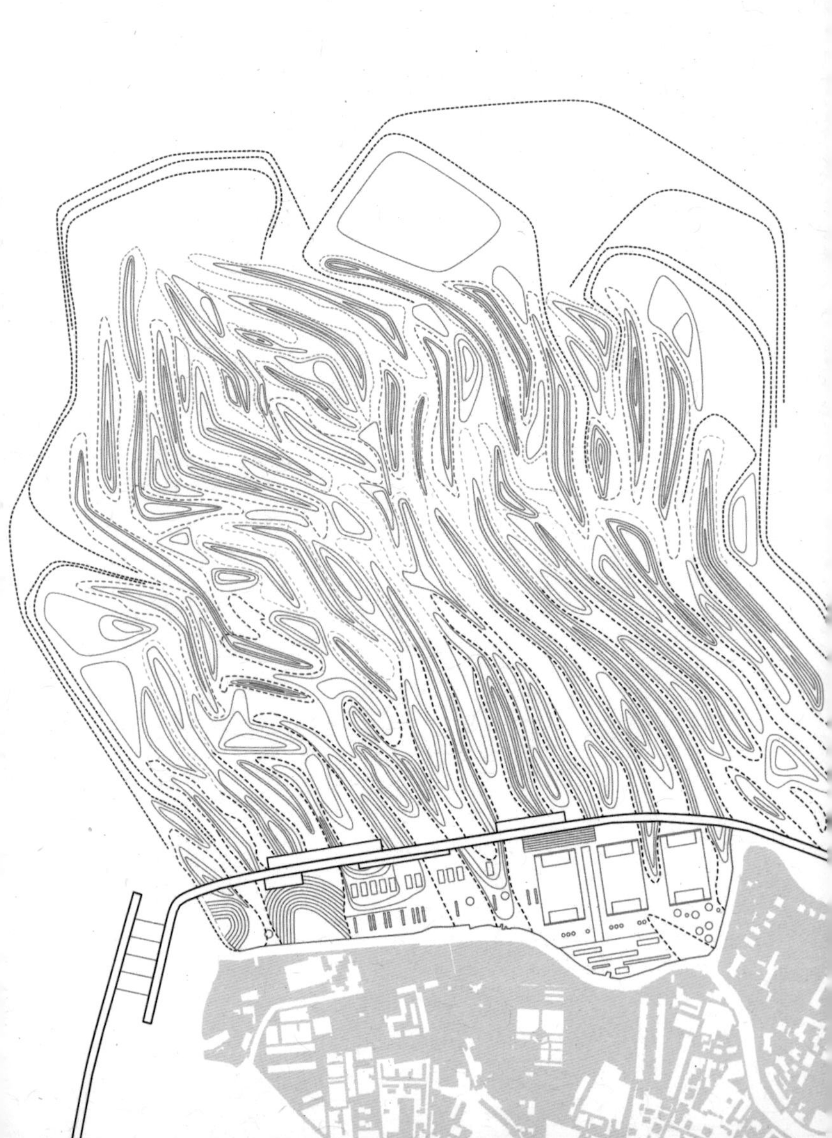

from the CO_2 in the carbon-rich atmosphere, thereby making the vessels net reducers of atmospheric greenhouse gases and largely self-sufficient in terms of food production.

As the new Venetian gondolas, the vessels will present an experience without parallel today. Travelers will be floating observers of powerful new technologies at peace with the terrifying power of the natural world. The Venice Lagoon Park will offer an optimistic vision of a world at once utopic and dystopic, a world in ruins and a world moving relentlessly into an unknown future.

CREDITS
Design drawings for this article are by RVTR (www.rvtr.com)
Acque Alte Team: Mark Friesner, Zhivka Hristova, Clayton Payer, Colin Ripley, Geoffrey Thün, Kathy Velikov.

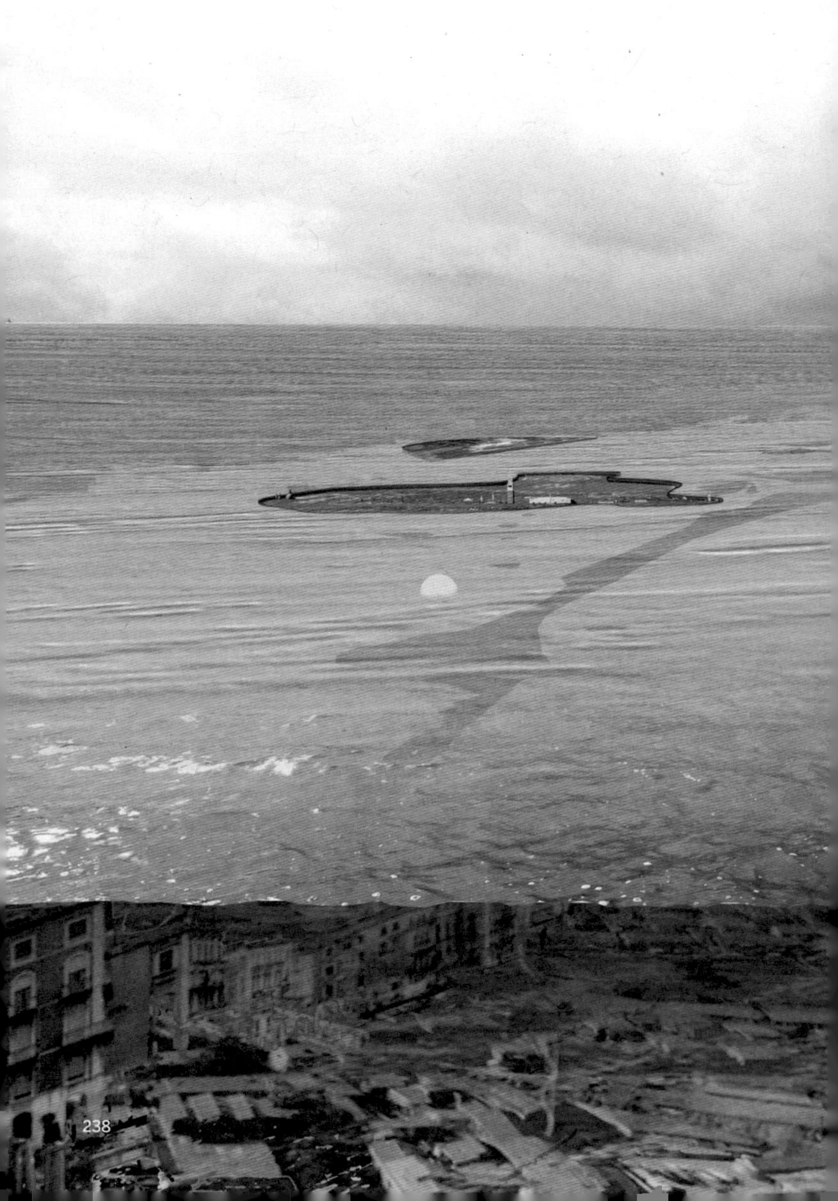

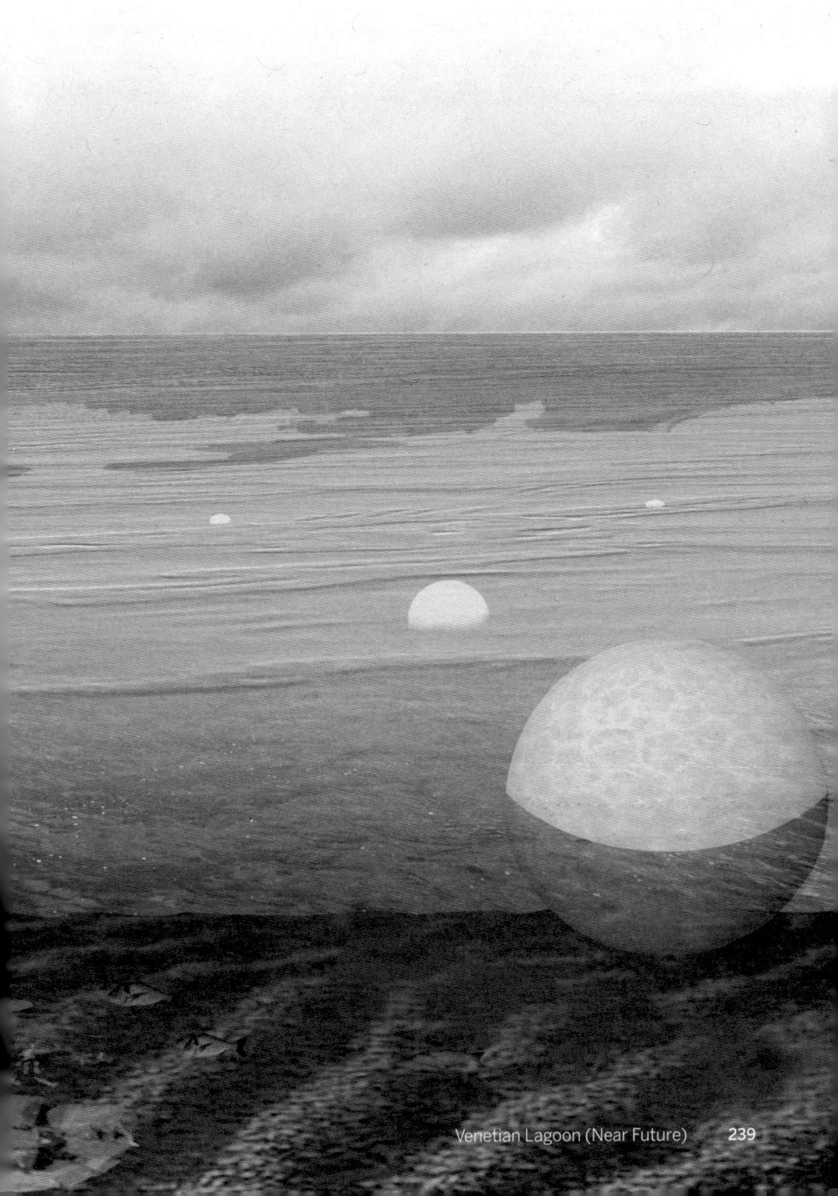

Contamination *Arnaud Maggs* My most recent work, "Contamination," shows pages of a water-damaged ledger from the 1905 Yukon Gold Rush. The factual transactions recorded in the book have been bypassed in favor of showing a series of mold formations that have migrated from page to page through the unused portion of the book. The soft, tainted pages and the bleeding pink ink are ghostly traces, phantoms of a cruel history, motivated largely by expectations of wealth. The spores are the animating principle of this work. The book itself is the host, the ground upon which these spores dance. As we approach the last pages of the book, we see the mold subsiding, gradually disappearing, leaving us with but a memory of what went before. The contaminated pages silently speak of a forgotten epoch of unfulfilled desires and delusional dreams...

Courtesy Susan Hobbs Gallery

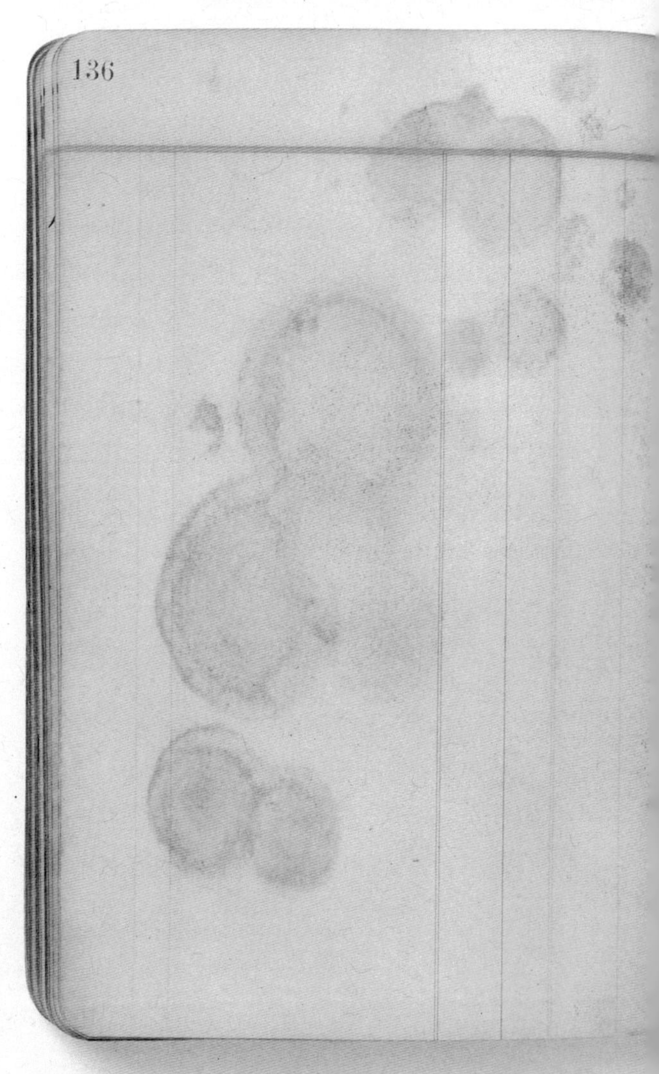

Contamination 136/137, 2007, color photograph, 2/3, 84 x 104 cm

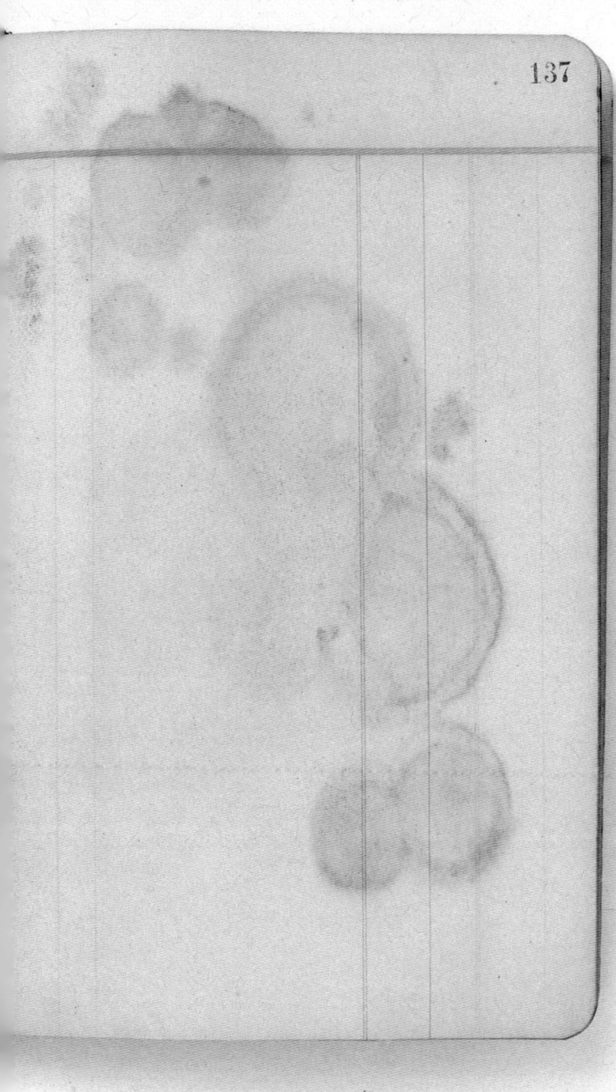

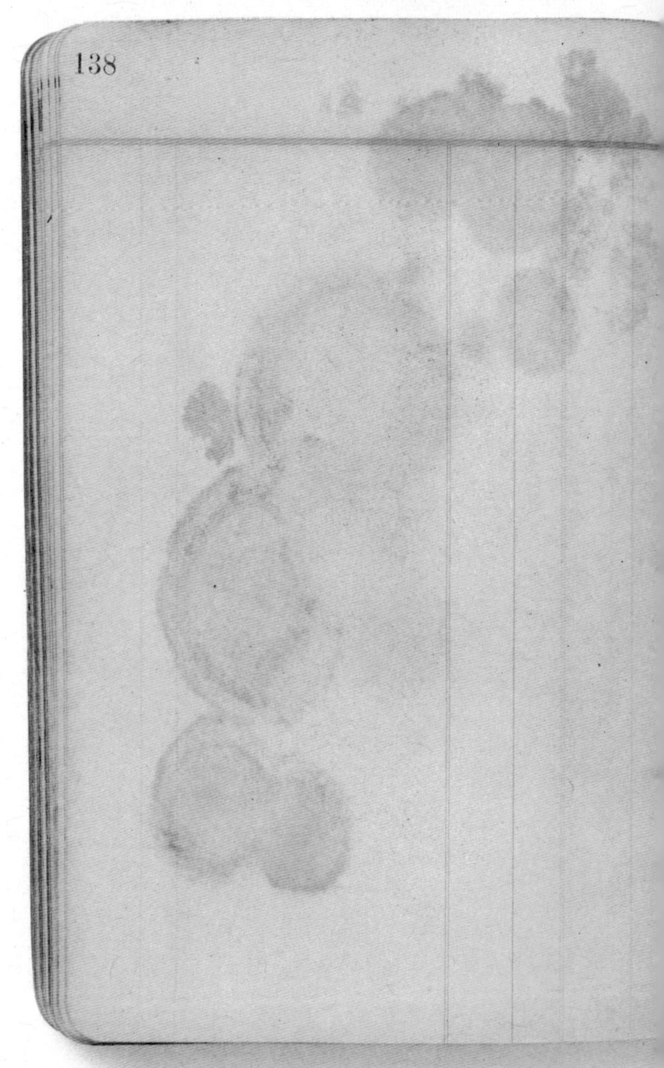

Contamination 138/139, 2007, color photograph, 2/3, 84 x 104 cm

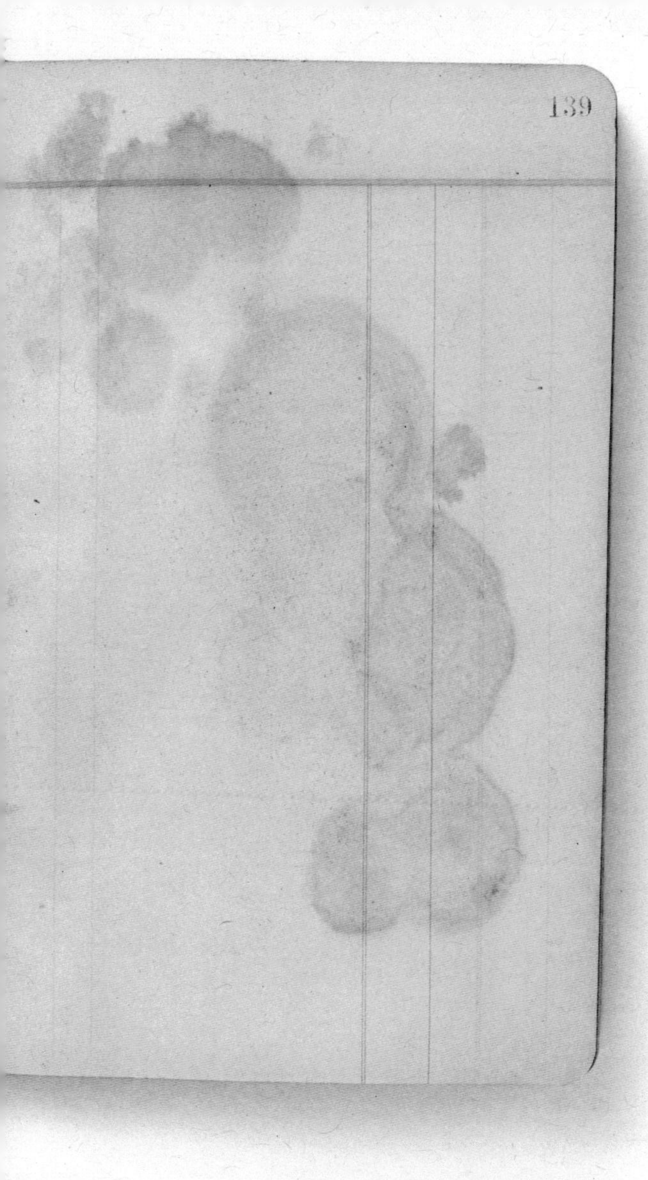

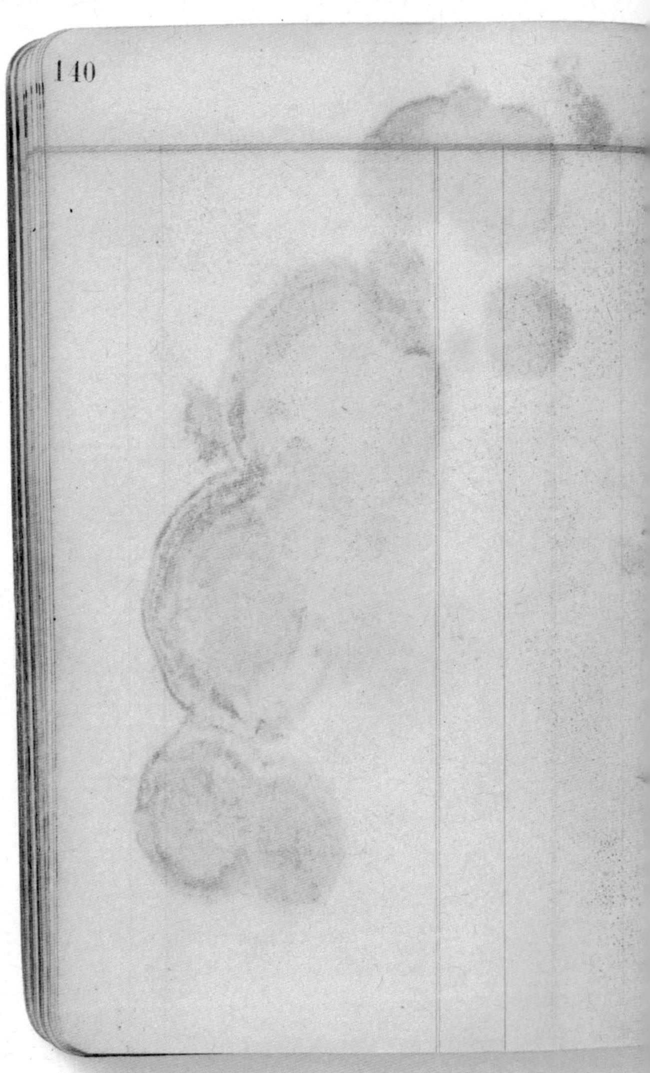

Contamination 140/141, 2007, color photograph, 2/3, 84 x 104 cm

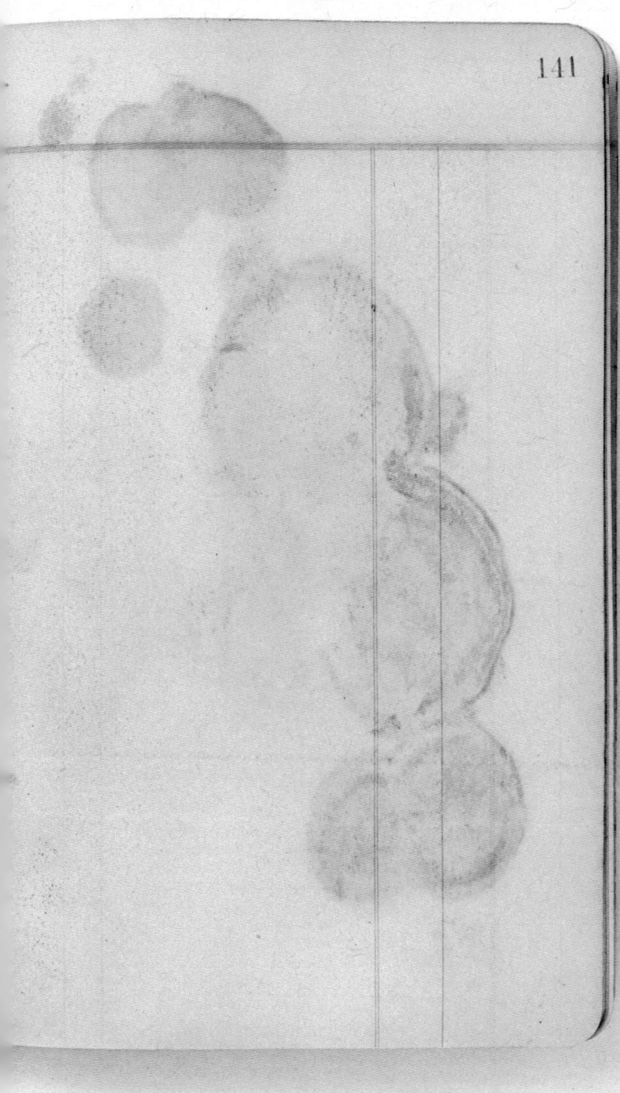

Contamination 142/143, 2007, color photograph, 2/3, 84 x 104 cm

143

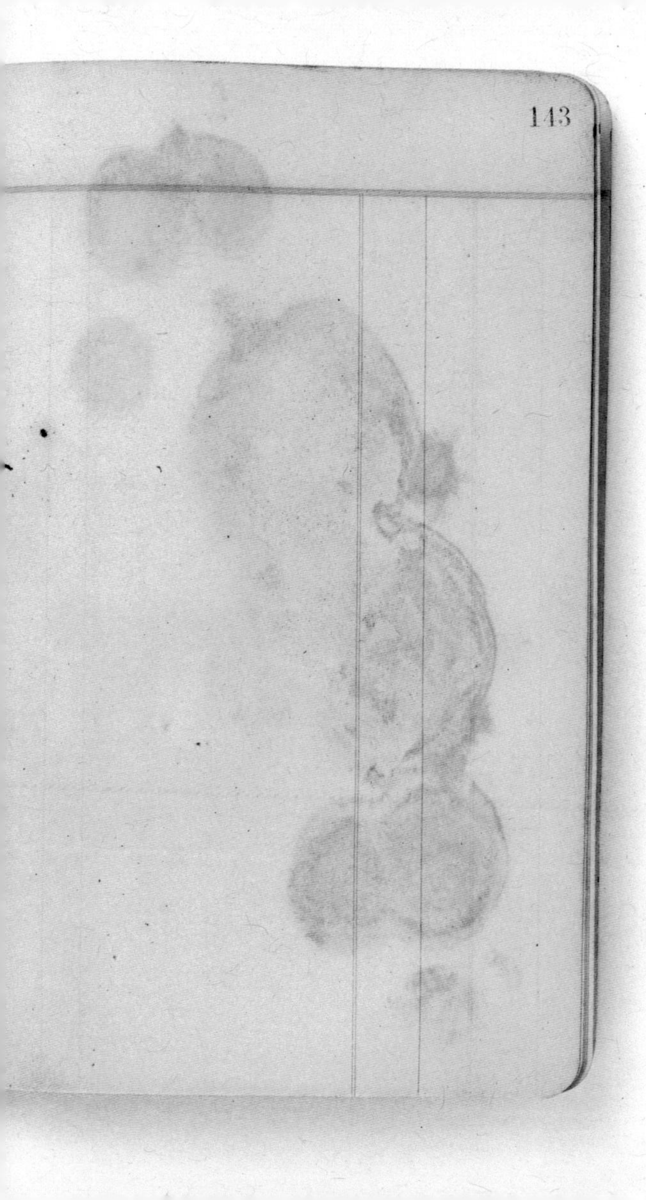

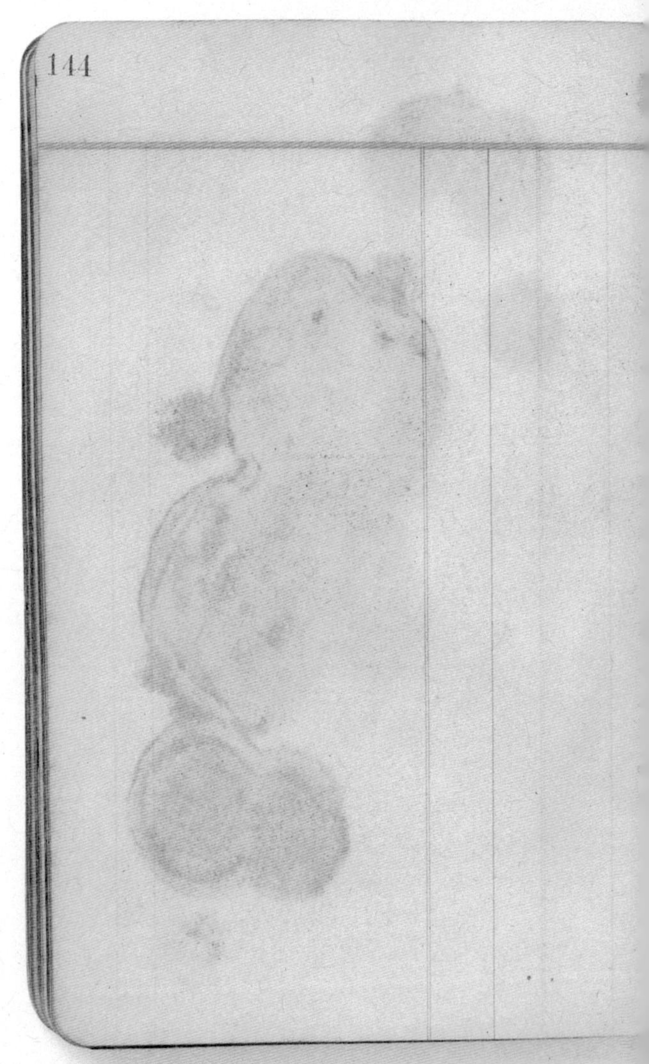

Contamination 144/145, 2007, color photograph, 2/3, 84 x 104 cm

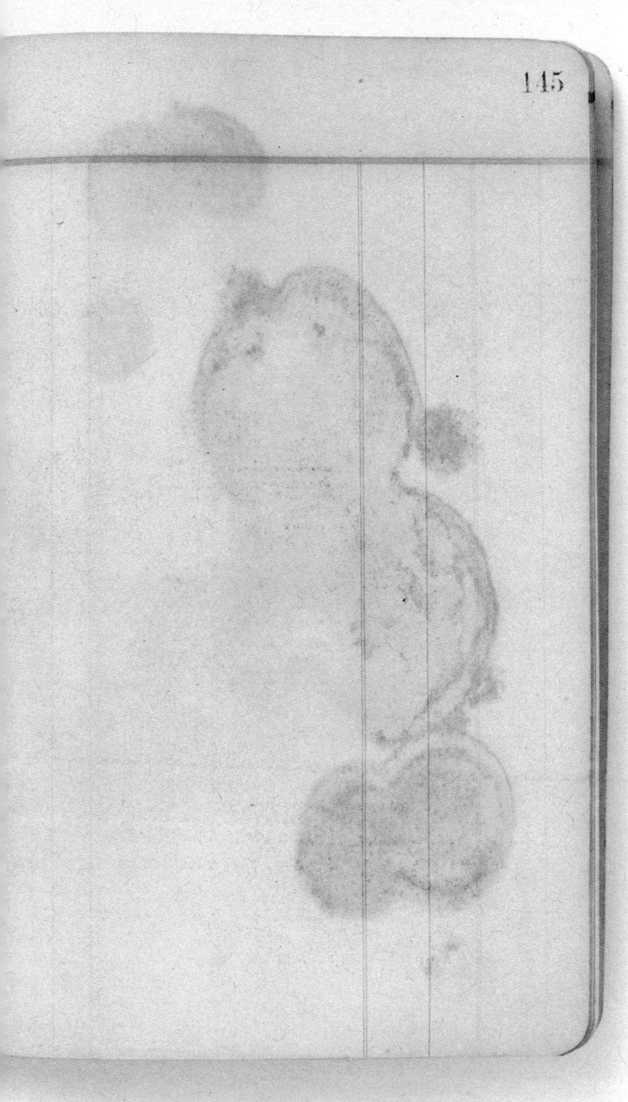
145

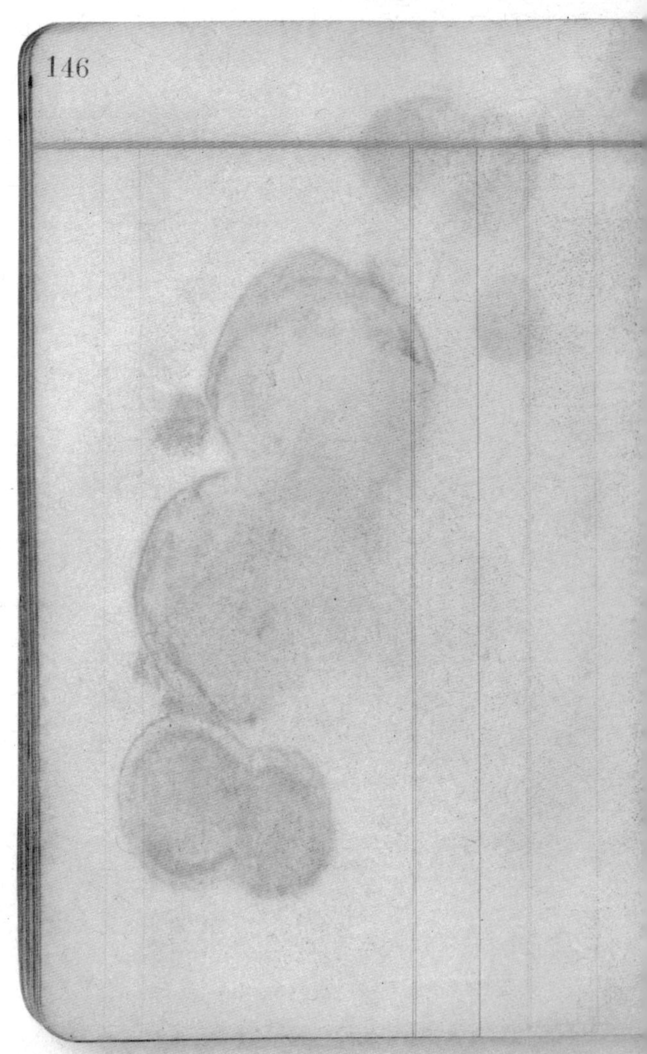

Contamination 146/147, 2007, color photograph, 2/3, 84 x 104 cm

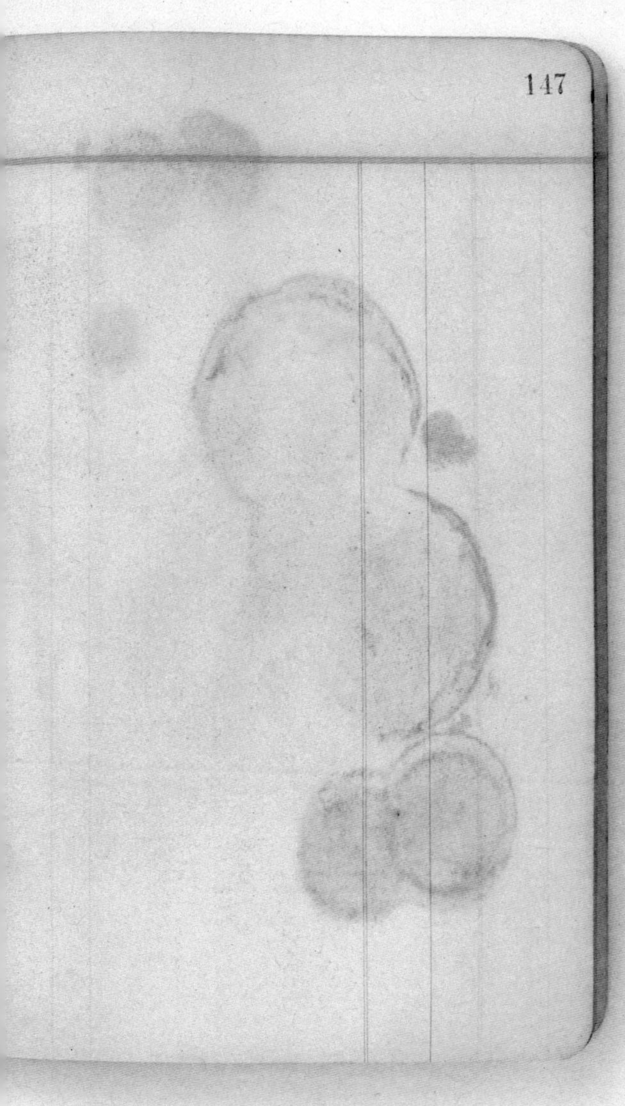

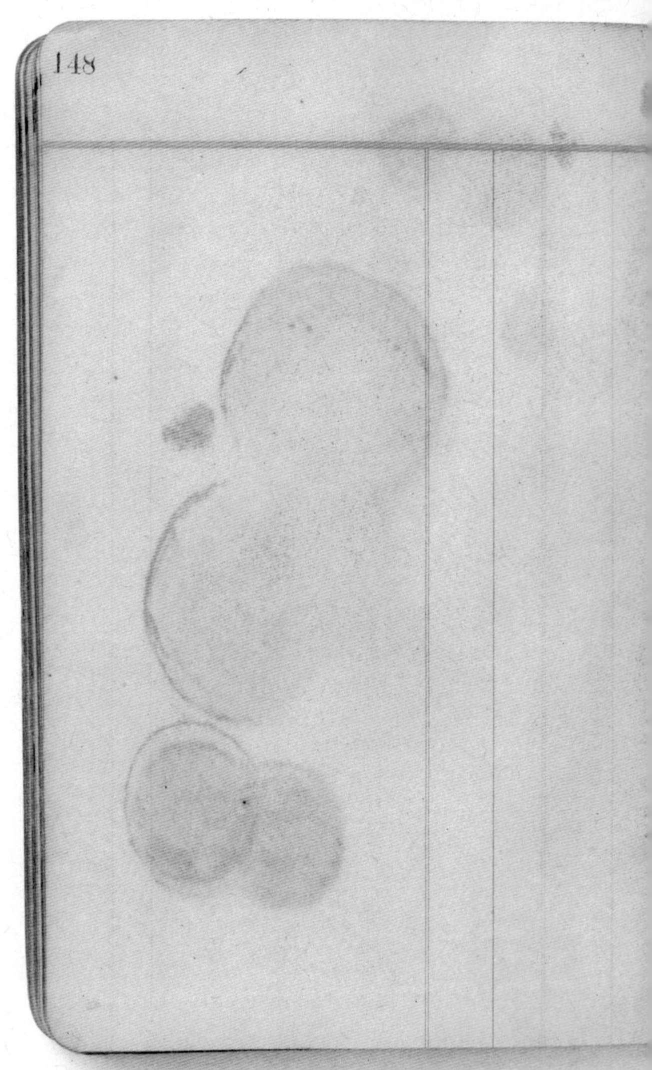

Contamination 148/149, 2007, color photograph, 2/3, 84 x 104 cm

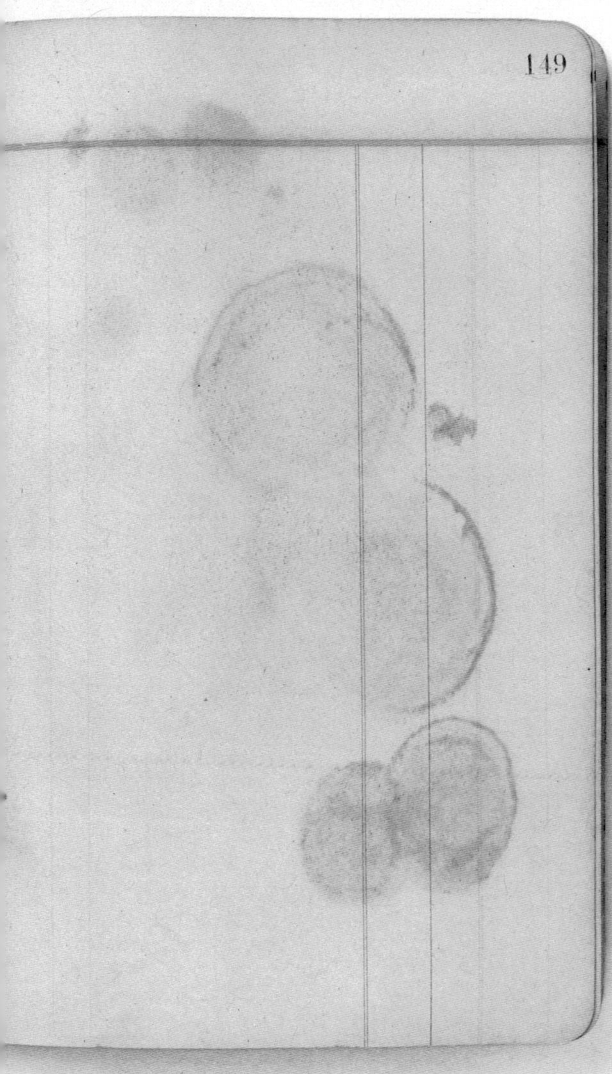
149

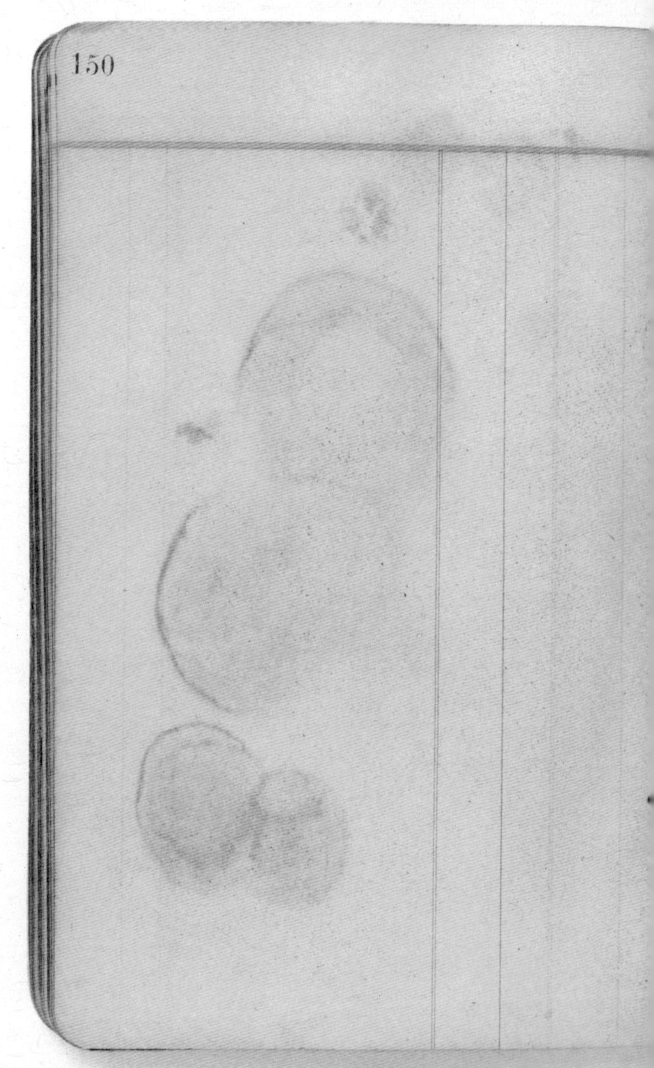

Contamination 150/151, 2007, color photograph, 2/3, 84 x 104 cm

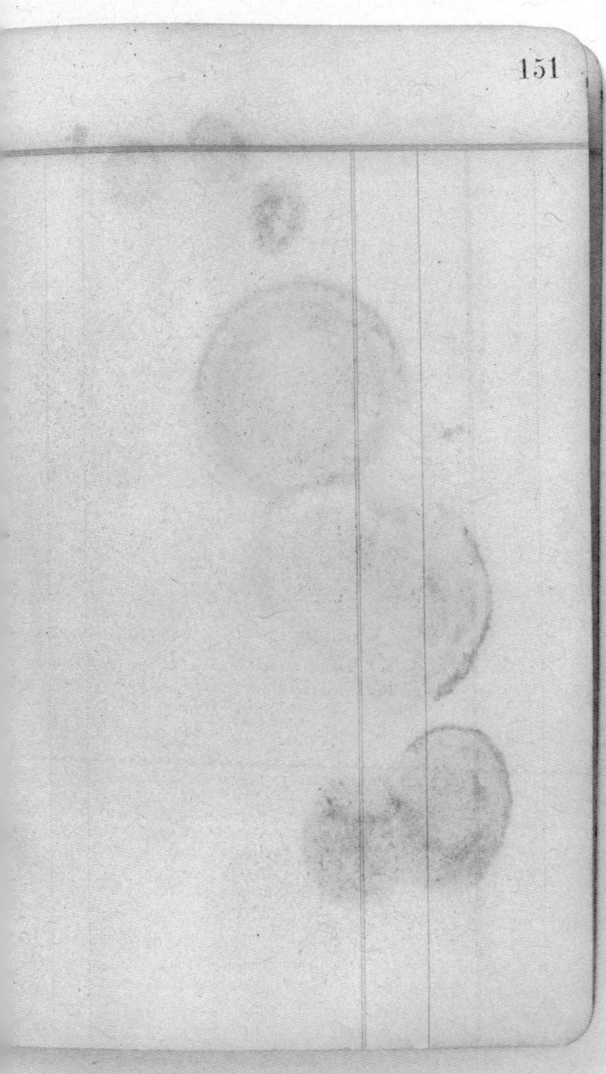

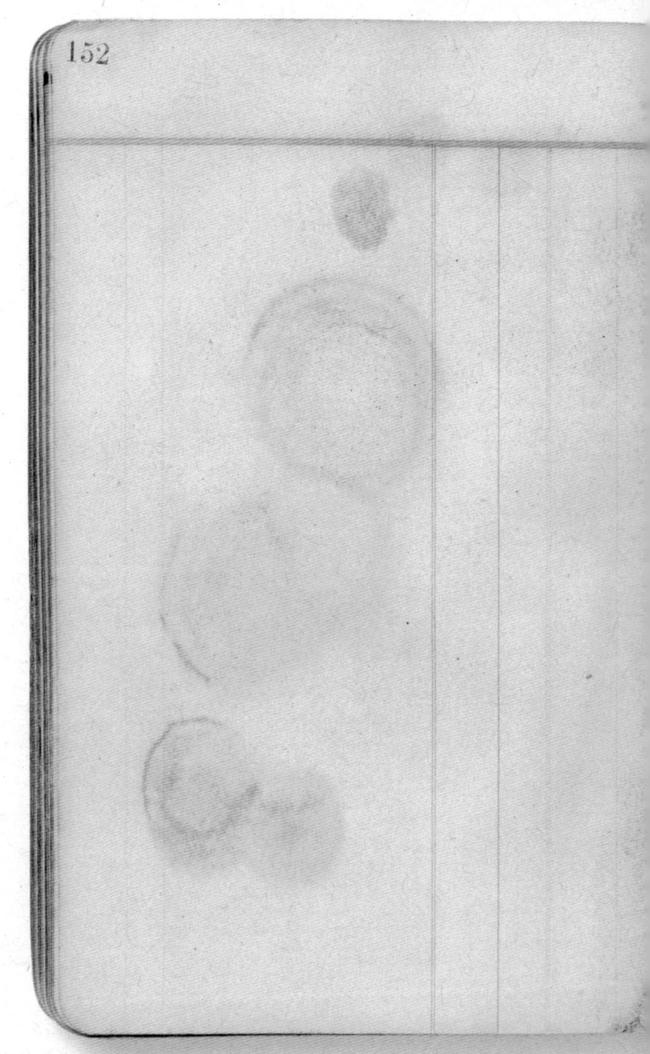

Contamination 152/153, 2007, color photograph, 2/3, 84 x 104 cm

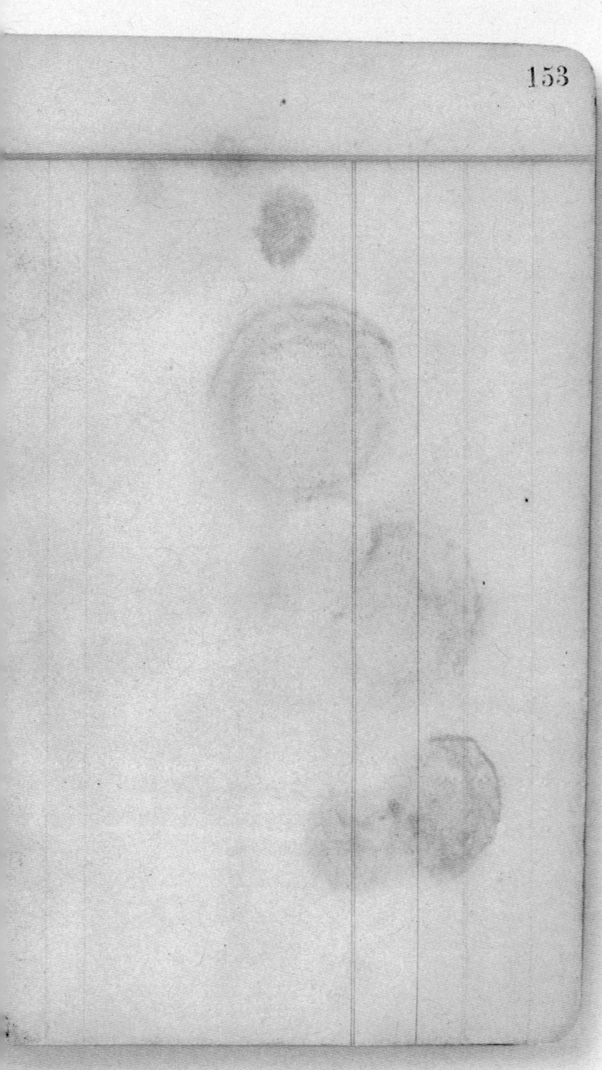

Contamination 154/155, 2007, color photograph, 2/3, 84 x 104 cm

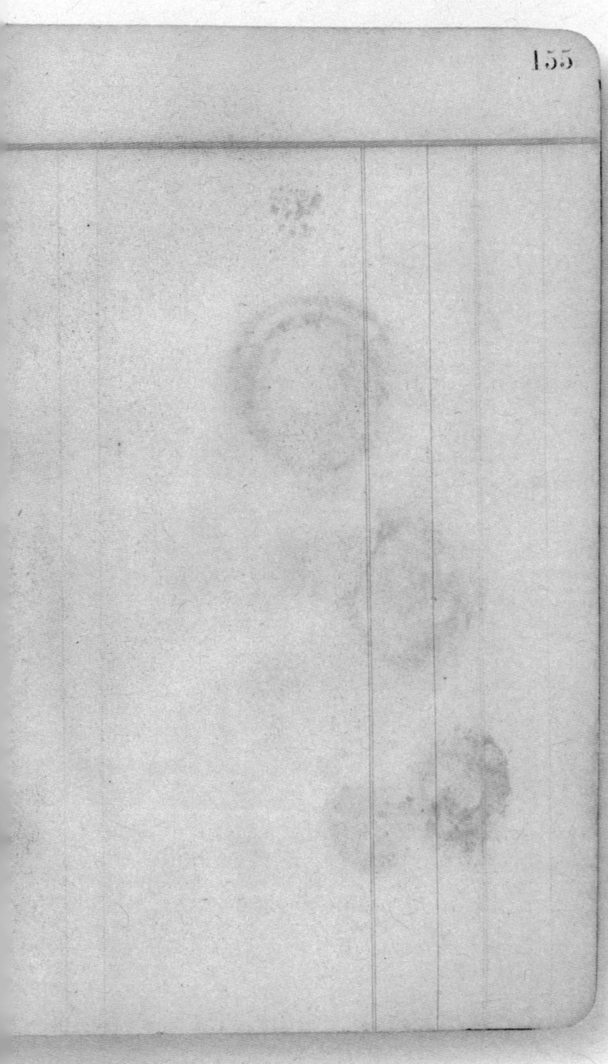

Contamination 156/157, 2007, color photograph, 2/3, 84 x 104 cm

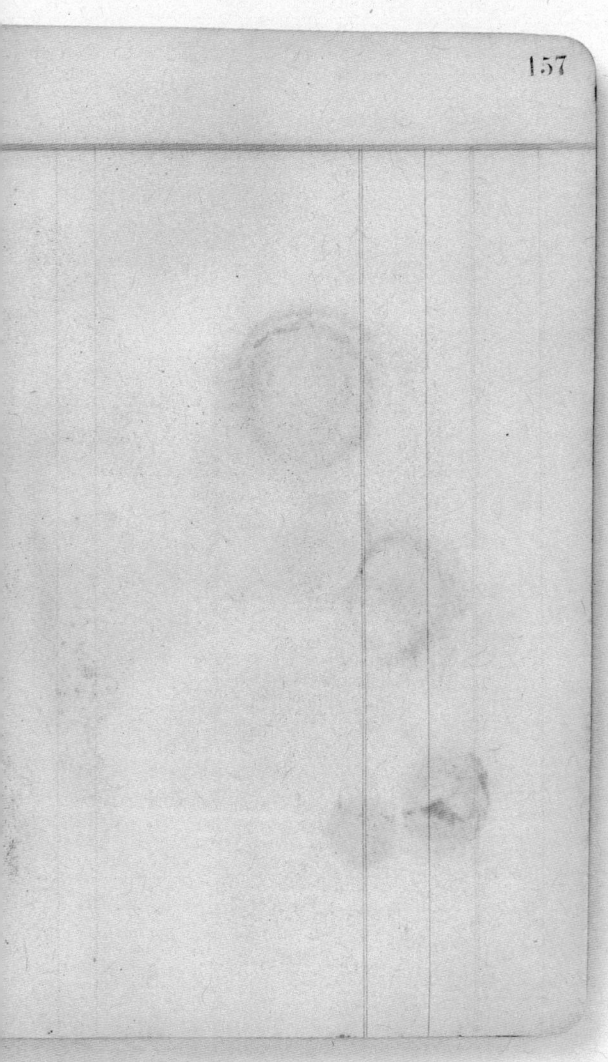

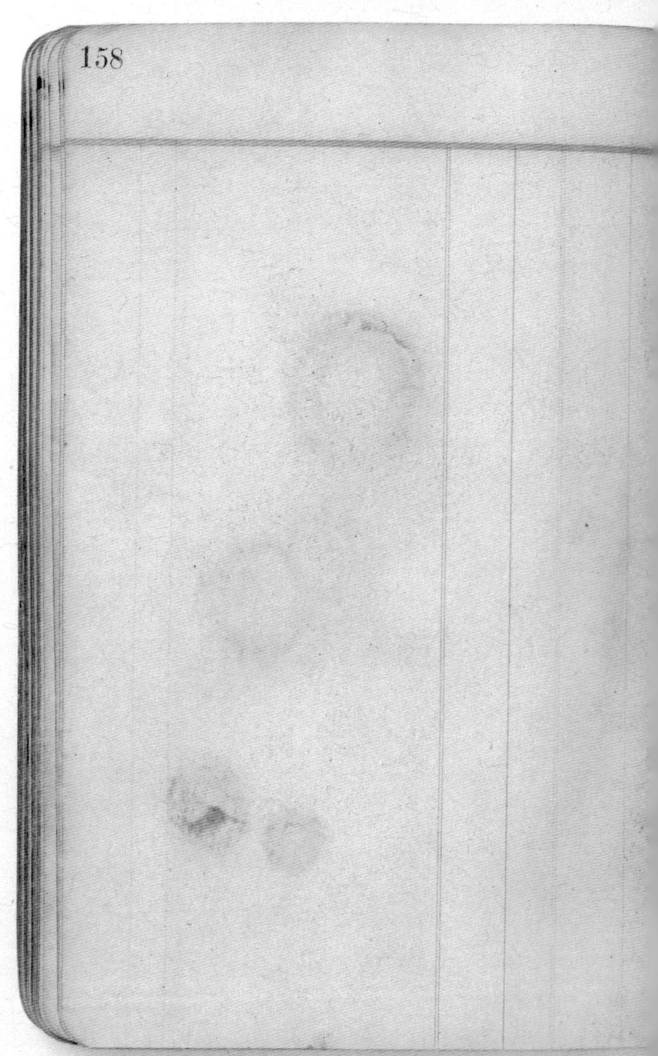

Contamination 158/159, 2007, color photograph, 2/3, 84 x 104 cm

Contamination 160/161, 2007, color photograph, 2/3, 84 x 104 cm

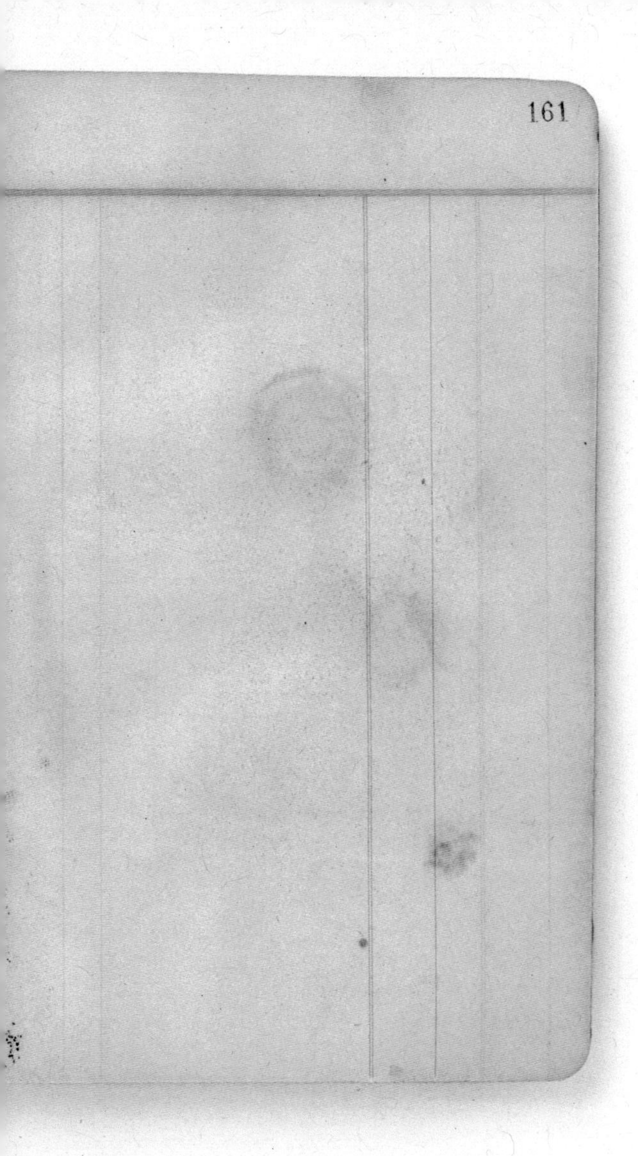

Contamination 162/136, 2007, color photograph, 2/3, 84 x 104 cm

Contamination 164/165, 2007, color photograph, 2/3, 84 x 104 cm

Contamination 166/167, 2007, color photograph, 2/3, 84 x 104 cm

Water, Inc. *Mei Chin*

Notes for a new campaign, WATER[1]

We at Water Inc. believe that literature could have been a better place if its characters had eschewed their usual beverages (mead, tea, absinthe) for WATER. It's a widely accepted fact that two liters of water a day is good for you, and being dehydrated can lead to fatigue, irritability, and hallucination. Therefore, if only those crazy characters that we love so much had drunk water they could have done away with depression, adultery, incest, violence, and pestilence.*

Literary Celebrities Who Did Not Drink Water include Emma Bovary, Gregor Samsa, Oedipus, and the Consul from *Under the Volcano*. Whereas Anne Shirley, of *Anne of Green Gables*, subsisted almost entirely on water drawn from the spring and the occasional glass of milk, and lived a healthy and happy existence. Aside from talking to trees and fairies, she did not exhibit any strange or morbid tendencies.

With this in mind, some of our most well-beloved characters ** have generously offered to demonstrate how their lives would have been different had they gotten their two liters of water a day.

* Maybe not pestilence. Simon from Marketing just pointed out that poor people died because of water-borne bacteria. Simon also suggests that we include the following: WARNING: Water is 92.8 percent plague-free, but if you experience boils, rashes, fever, or death you should consult your physician immediately.

** on loan from the Union of Tragic Literary Personalities UTLP

1 NOTE: Must not repeat the debacle of last year's campaign: SLEEP! Salvador Dali didn't, but Calvin Coolidge got plenty.

Dear Water Inc.

I have swapped out all the Rhine draught for water and have noticed such an improvement!

Since I was drinking water instead of wine on the parapets, I never saw the ghost of my father. We retired at about five in the morning—cock's crow, so we were pretty sure that there would be no ghosts after that—for tea and toast and it was pleasant catching up with the boys.

The rest of the court is drinking water too! Ophelia, instead of drowning in the brook, realized that the water was full of minerals—she floated a penny on its surface! So she bottled it and started to sell it to specialty stores. We are all so proud of our fair maid! Next year she will be opening a chain of water and oxygen bars, but for now you can get her product on www.lethesbramble.com.

Mother never drank from the poisoned chalice of wine. I mean, who drinks wine when you're in the middle of a life-and-death fencing match? She found the water a refreshing alternative.

Horatio mentioned that the wine was making my flesh a little too solid, and that the water has made a great improvement on my figure. I am no longer fat and scant of breath.

Also I am proud to say that I have graduated with honors from Wittenberg (finally! At the age of thirty!) and have agreed to do a series of motivational talks around Europe. My first book, *Get Over It!* is out in bookstores, and my second, *It's not your mother's fault*, is due out next year.

You can find more information about me on my website: www.notyourmothersfault.com.

Gratefully,
Hamlet, Prince of Demmark.

Franz Kafka comes home from a long day at the insurance company, writes "Metamorphosis" on the page, and then, realizing he is thirsty, goes into the kitchen. He hesitates over coffee, absinthe, Ovaltine, beer, and then finally he lights on a six-pack of water. He drinks the water, writes "METAMORPHOSIS" on the title page, then calls his mother, calls his two fiancées, lifts weights, and falls asleep. Last night he dreamt of a man starving in a cage; the night before that, he dreamed of being blown away in a bucket, but tonight he dreams of models cavorting half naked, scampering around a lake. The next day he wakes up, feeling curiously refreshed, and writes his first sentence, "When Gregor Samsa woke up, he realized that something was different. He resolved to eat more fresh fruits and vegetables, quit his job as a salesman, and pursue his life-long dream of inventing a talking toothbrush."

Today Mother died. Or maybe it was yesterday, I am not sure. Which made me think that instead of coffee or brandy I should probably have water. I tasted it and it didn't taste bad. So I flirted with the Pilates instructor and got her number, and then went to my lawyers, sorted out the business with mother's will, drank more water, made an egg white omelet and realized that I had really had a very productive day.

Anna Karenin, or Anna K. as she prefers to be called now, declined to submit anything but agreed to an interview. "I'm so incoherent and unfocused and rambling..." she explained. "Anything I give you is bound to be too long" —she paused playfully—"by about five hundred pages."

She arrives looking smaller and more girlish than one might expect, but her soft footstep is hers, as is the milky skin, and her black hair fitting her like a snood.

Tentatively, we remind her of her last thoughts before she throws herself under the train:

"Yes, I'm very much worried, and that's why reason was given to me, to escape; so then one must escape: why not put out the light when there's nothing more to look at, when it's sickening to look at it all? But how? Why did the conductor run along the footboard, why are they shrieking, those young men in that train? Why are they talking, why are they laughing? It's all falsehood, all lying, all humbug, all cruelty!..."

Anna giggles and claps her hand to her throat.

"My goodness," she said. "That was me, wasn't it. Whoa! That was the result of too much black coffee. That's all I was living on—black coffee and," she lowers her lashes, "...and painkillers. Can't you tell?"

"You know," she says, settling more comfortably into her chair and sipping her water thoughtfully, "I never liked him, because he looked down on me, but I really should have taken a page out of Levin's book. I mean, he was a water drinker, made him one with the peasants and all that. Clean air, exercise in the fields, cabbage soup diet, and always water, water, water." She shrugs. "I should not have had those last four coffees after leaving my brother's house. You think your heart is breaking, when really it is just overstimulated by the caffeine."

She turns to greet a visitor who has just entered the room—with a similar soft footstep, pale complexion, and black hair. "Emma!" she says, extending her hand. "Chérie!"

"Anna," says the visitor, "chérie! I am not interrupting anything at the moment, I hope?"

"No," says Anna K. with a low laugh. "Please join us." They kiss each other on the cheek.

The lady known as Emma says, "Oh, I did receive your letter, but you know, apologies that I did not respond, but you know, I only read romantic novels and misbehave." She spreads her white fingers with their oval-shaped fingernails, as if to say, are these hands that work?

"But," says Emma Bovary eagerly, "this eau, it has been a révélation. I have realized only too late that Rodolphe was getting a pot belly and that Leon was insupportable, but more than that—if you drink coffee and wine and then have sex and then are expected to spend your afternoons in some fashion, what do you do? You shop! And you shop, shop, shop, and you have no idea how much money you are spending because you are high from the wine and energetic from the coffee, and than, mon dieu, in the evening, you are listless—"

Anna finishes, "—and just want to kill yourself."

Emma nods. The water has brought a healthy flush to her cheeks and a sparkle to her dove-gray eyes.

Anna says, "And even after all that water, if I still wanted to kill myself, I would have found a less dramatic, but less painful way. I mean, honestly, a train? I know that it gives the novel *Anna Karenina* structure, but a train."

Emma fans herself, "I mean, chérie, the train, it is so excessive when you have all that morphine at home, no? and then I am married to a doctor, surely, there is a more intelligent alternative than arsenic."

What alternative is their favorite?

The ladies laugh.

"We call it," Anna says, "the Lily Bart cocktail. Half a bottle of chloral, one glass of heavenly, pure, sweet spring water, stir, drink and sleep."

Socrates: "I'm about to drink hemlock and die. Do you think I want to be sober? P.S. Plato says that thanks to your 'product' he has no dialogues."

Lady Brett Ashley: "I don't know, chaps. Michael didn't fight with Cohen, I didn't have sex with the delicious matador, and we talked earnestly about yoga and how the water 'was cold. And good. And it was cold and good in our mouths on a hot day.' Do you know how awful that is when you're in Pamplona? I was so disgusted I drank two liters of vino clarete and passed out."

Jesus Christ: "At the Marriage of Cana, I changed water into water, rather than into wine and died prematurely."

ENDNOTE: Not all responses from the Union of Tragic Literary Personalities have been as promising.

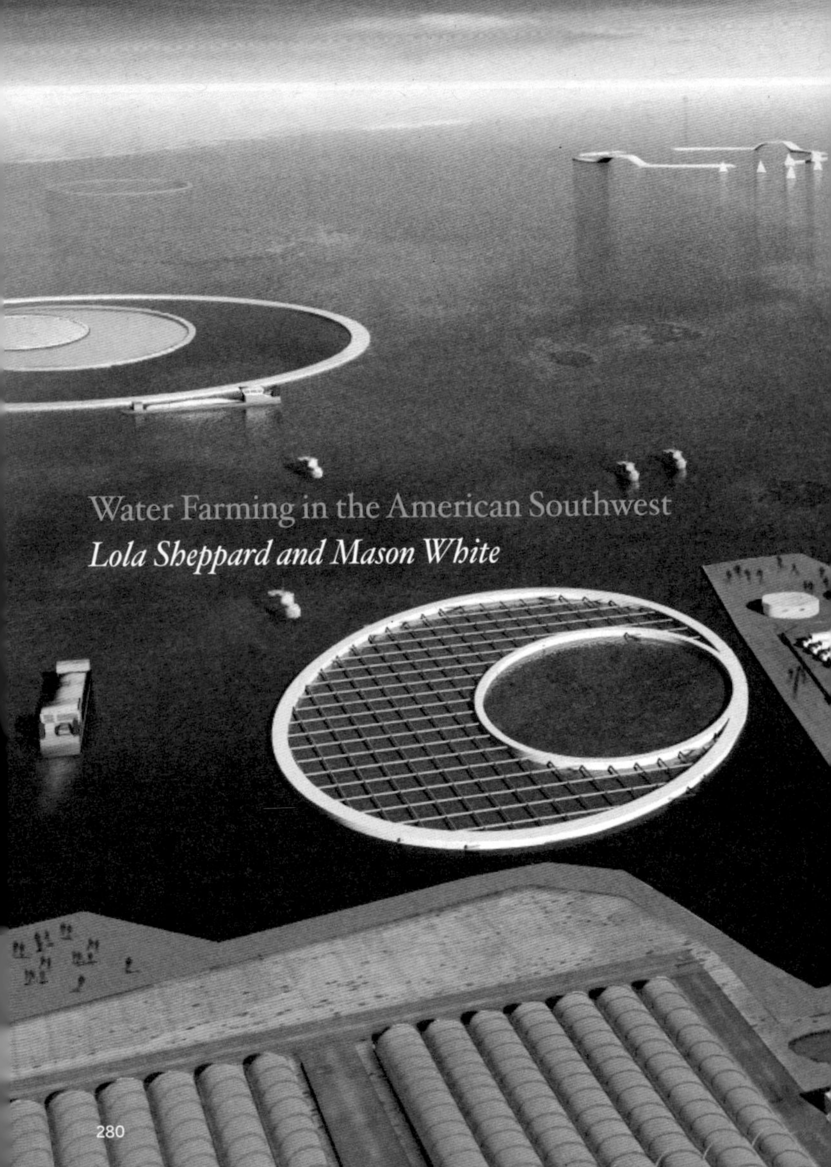

Water Farming in the American Southwest

Lola Sheppard and Mason White

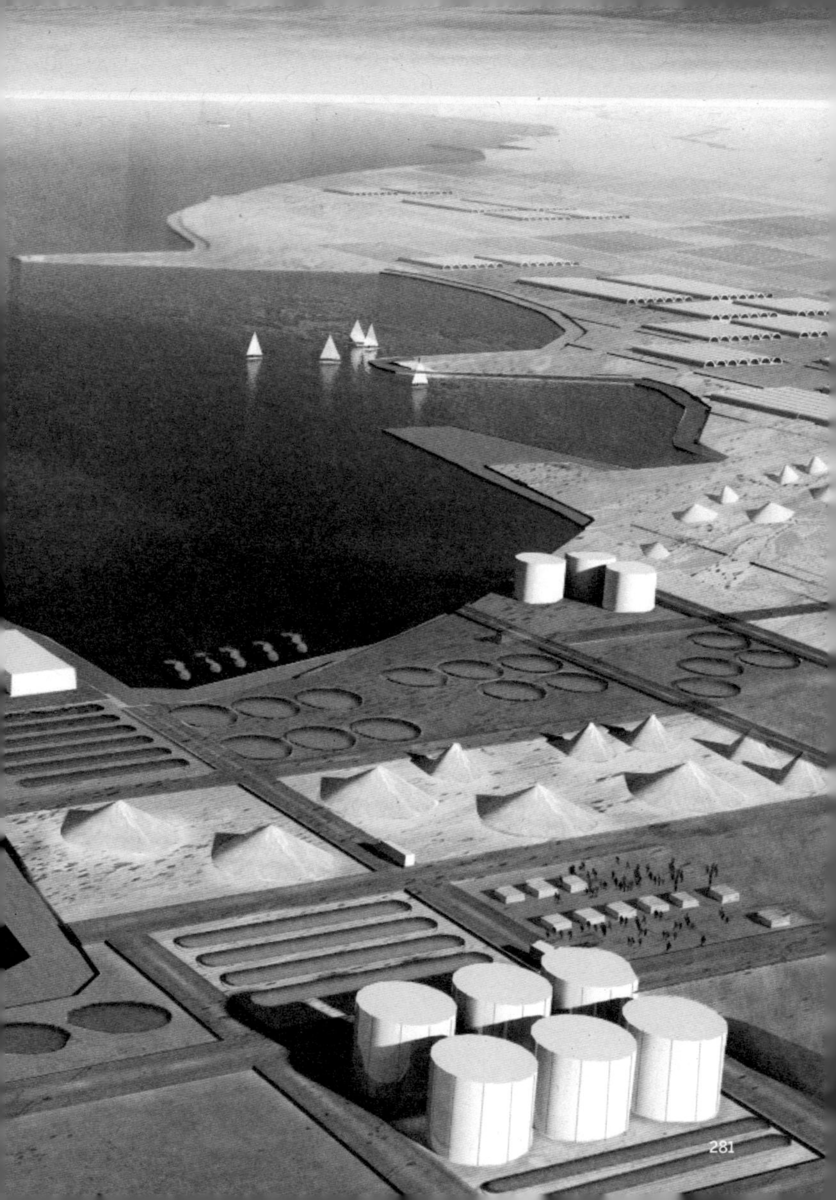

The United States has 6.5 percent of the world's fresh water within its borders, yet despite this abundance the country is entering a period of crippling water shortages. Even without factoring in the expected drought effects of climate change, thirty-six states predict shortfalls in the next ten years that will put their environmental and economic health at serious risk. Making the problem more acute, the driest states are also the fastest growing: six of the ten fastest-growing cities in the US—Dallas/Forth Worth, Phoenix, Houston, San Bernardino, Houston, and Las Vegas—are located in water-strapped regions of the Southwest. The constant drive for urban growth in the face of water scarcity necessitates ever more elaborate infrastructure to transport water from wet regions to dry ones. Heroic efforts in the American Southwest demonstrate what can be achieved, and at what cost, when this essential resource must be supplied to communities that have no ready access to water.

INFRASTRUCTURAL LANDSCAPES

Man-made water networks have successfully enabled cities to grow and prosper in the desert conditions of the Southwest, but impending water shortages mean that increasingly complex regional water infrastructure is urgently needed, especially in the 243,000 km2 catchment area of the Colorado River basin.

The Colorado River is the site of ambitious feats of American engineering, many initiated during the 1930s under Franklin D. Roosevelt as part of the New Deal. Most notable is the Hoover Dam on the border between Nevada and Arizona, which created Lake Mead, the largest man-made lake in North America. The dams along the Colorado River collect water in enormous reservoirs, thus ensuring consistent supplies throughout the year, managing river levels, and providing hydroelectric power to the region. The reservoirs—which store more than 86 billion cubic meters of water, or four times the Colorado River's average annual volume—feed an artificial network of canals and pipelines. Three major aqueducts, essentially man-made rivers, divert water in the lower Colorado River basin. Combined, they form approximately 1065 km of water highways.

In the years between 1935 and 1993 almost every mile of the Colorado River was modified in some way by this system of dams, reservoirs, aqueducts, and pumping stations. Most of the river's flow in normal hydrologic years is now diverted for agricultural and municipal water supply, leaving very little to reach its mouth in the Pacific. In its present form, the basin's water management system has evolved into a symbiotic network between cities, landscape ecologies, and infrastructure technologies.

PARTITIONING WATERS

Even in the early days of development in the Southwest, water was a critical issue. As population centers grew and the attendant demand for water became ever more pressing, states dependent on Colorado River water feared California would establish priority rights. The 1922 Colorado River Compact was designed to permanently resolve this conflict. The agreement divided the waters of the Colorado River equally between "Upper and Lower Basin States." Wyoming, Colorado, Utah, and New Mexico were in the upper basin; and California, Arizona, and Nevada in the lower, with the demarcation line set at Lee's Ferry in northern Arizona. Each basin was to receive 7.5 million acre-feet per year (293 m^3/s). It later became apparent that these allocations, based on hydrological data from a period when the Colorado River's flow was unusually high, would not be sustainable over the long term.

In 1963, a US Supreme Court decision redefined the agreement, reducing the quantity of water that was to be distributed among the lower-basin states, and revising upward the amounts that had been historically reserved for Indian tribes and federal public lands. Because the Colorado River Delta is in Mexico, that country also has a legitimate claim to its waters. The delta was once lush with vegetation and wildlife, but the construction of 29 dams and numerous up-river

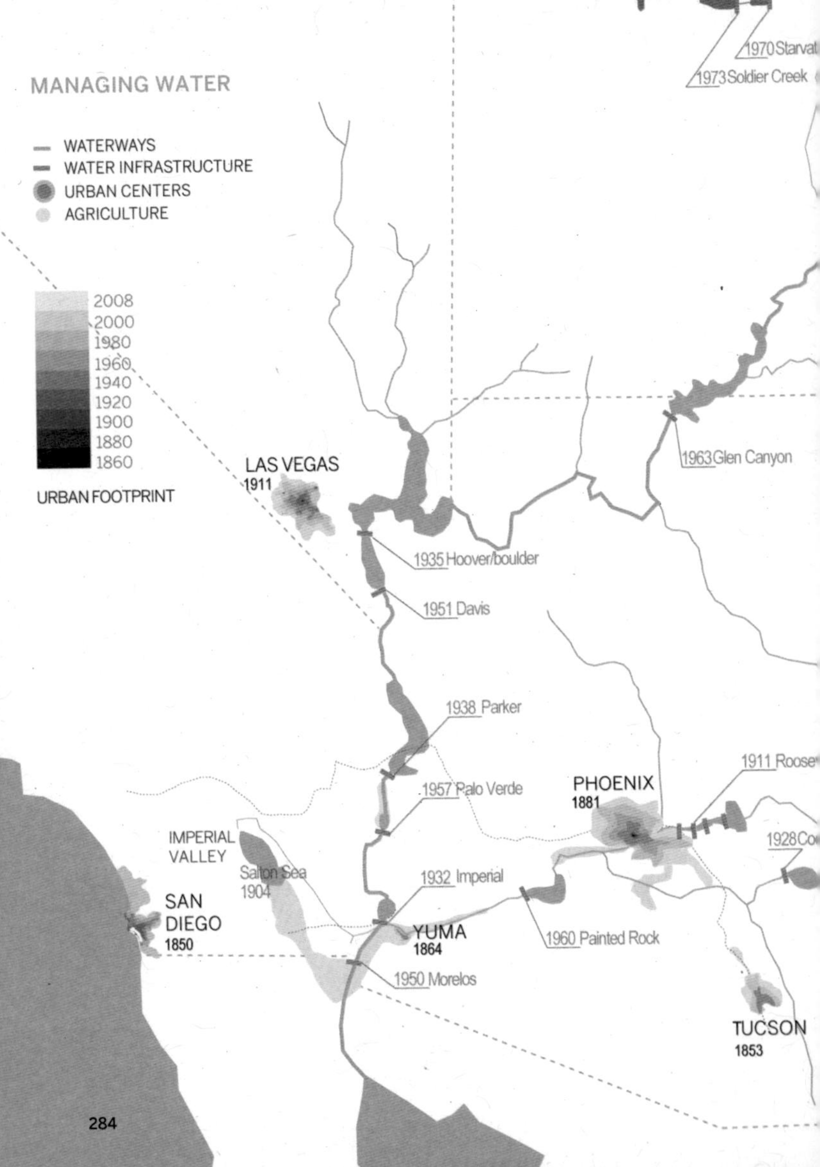
MANAGING WATER
WATERWAYS
WATER INFRASTRUCTURE
URBAN CENTERS
AGRICULTURE
2008
2000
1980
1960
1940
1920
1900
1880
1860
URBAN FOOTPRINT
1970 Starvat
1973 Soldier Creek
LAS VEGAS
1911
1935 Hoover/boulder
1951 Davis
1963 Glen Canyon
1938 Parker
1957 Palo Verde
1911 Roose
PHOENIX
1881
1928 Co
IMPERIAL
VALLEY
Salton Sea
1904
1932 Imperial
SAN
DIEGO
1850
YUMA
1864
1960 Painted Rock
1950 Morelos
TUCSON
1853

diversion projects over the past 60 years deprived the area of natural water flow, with its vital supply of silt and nutrients. In 1944, an international treaty allocated Mexico water rights in the amount of 58.6 m3/s. In the 1960s, Mexican farmers complained that when the untreated runoff from the US made its way into the Colorado River as it headed south of the border, it raised the salinity levels, harming their crops. In 1992 one of the world's largest desalination plants was built in Yuma, Arizona, to treat agricultural runoff before it was delivered to Mexico. The competition for water in the Colorado River Basin continues to be an engineering and political problem. Given projected growth in the region, these interstate and international debates are only likely to intensify.

WATER TRADING / WATER BANKING

Every US state is divided into irrigation districts, legal bodies that trade in water capital. These districts are quasi-political municipal or regional entities created under special laws, for the purpose of supplying water and power to agricultural regions and cities. Farming districts are now selling their long-established water rights to thirsty urban centers. Water has become a highly valuable, tradable commodity. For example, a deal struck in summer 2008 between the Metropolitan Water District of California (MWD) and the Paolo Verde Valley called for farmers to divert 4.5 m3/s m^3/s of their water to Los Angeles, San Diego, and surrounding settlements in exchange for US$16.8 million a year. This desperate and expensive negotiation was necessary because the management of the Sacramento River Delta had reduced its contribution to the MWD by 30 percent, to protect a threatened fish species.

The legal frameworks regulating land and water rights in the western US are known as "first in time, first in right" or "prior appropriation" laws. Under these rules, water rights can be severed from the land, and sold or mortgaged as an independent piece of property by those who had the original, historical claim to the land. This is a

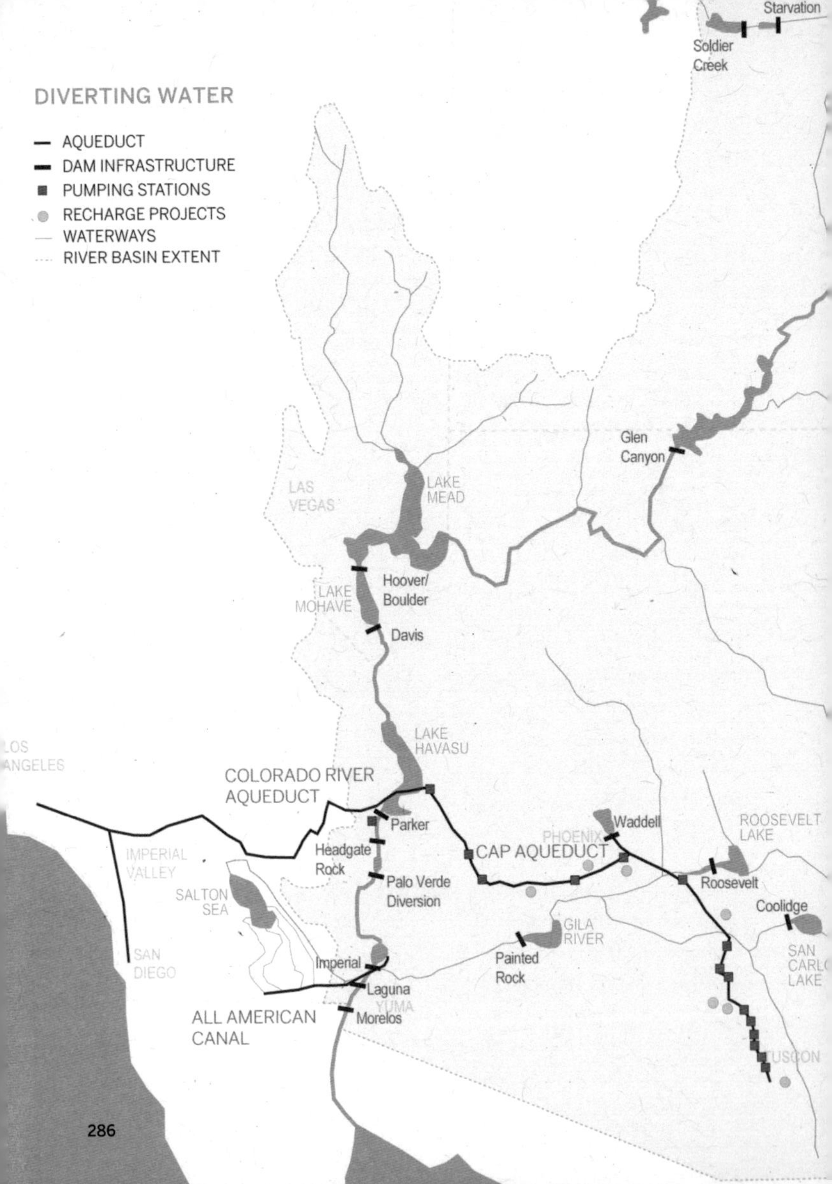

DIVERTING WATER
AQUEDUCT
DAM INFRASTRUCTURE
PUMPING STATIONS
RECHARGE PROJECTS
WATERWAYS
RIVER BASIN EXTENT
Starvation
Soldier Creek
Glen Canyon
LAS VEGAS
LAKE MEAD
LAKE MOHAVE
Hoover/ Boulder
Davis
LAKE HAVASU
LOS ANGELES
COLORADO RIVER AQUEDUCT
Parker
Headgate Rock
Palo Verde Diversion
IMPERIAL VALLEY
SALTON SEA
SAN DIEGO
Imperial
Laguna
YUMA
Morelos
ALL AMERICAN CANAL
PHOENIX
Waddell
CAP AQUEDUCT
ROOSEVELT LAKE
Roosevelt
Coolidge
SAN CARLO LAKE
GILA RIVER
Painted Rock
TUSCON

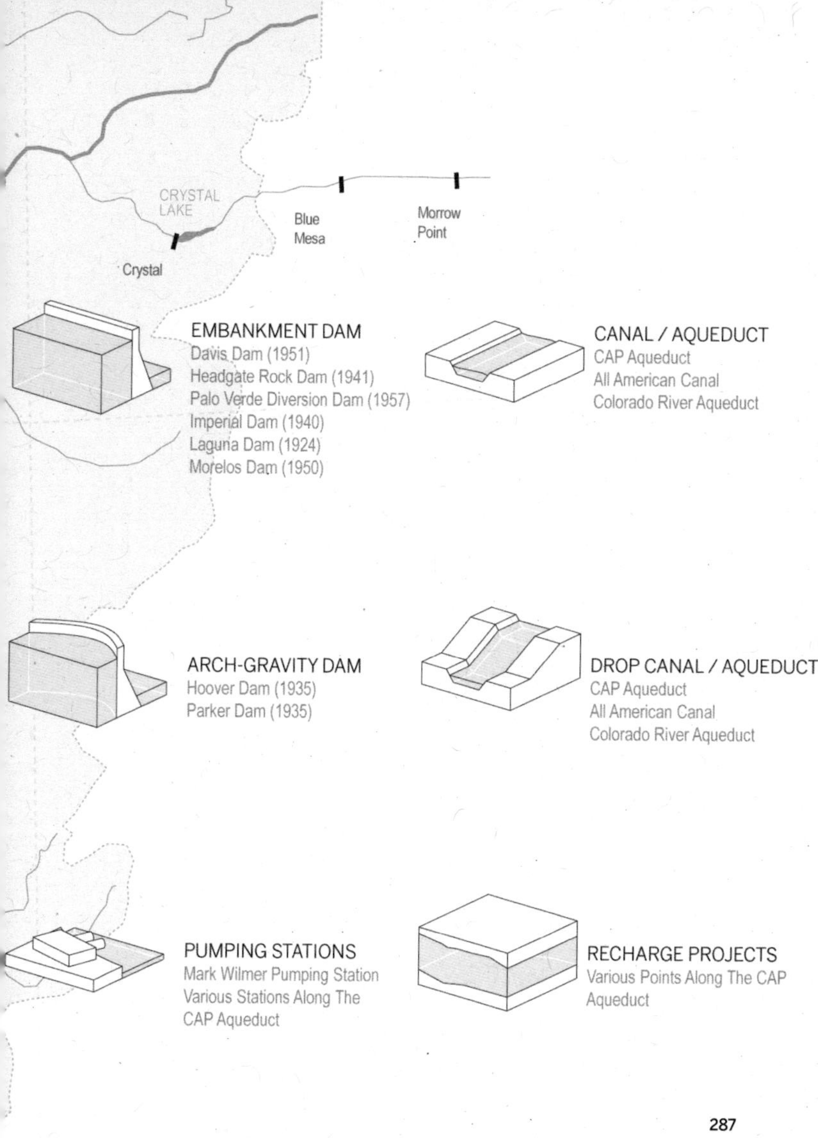
CRYSTAL
LAKE
Blue
Mesa
Morrow
Point
Crystal
EMBANKMENT DAM
Davis Dam (1951)
Headgate Rock Dam (1941)
Palo Verde Diversion Dam (1957)
Imperial Dam (1940)
Laguna Dam (1924)
Morelos Dam (1950)
CANAL / AQUEDUCT
CAP Aqueduct
All American Canal
Colorado River Aqueduct
ARCH-GRAVITY DAM
Hoover Dam (1935)
Parker Dam (1935)
DROP CANAL / AQUEDUCT
CAP Aqueduct
All American Canal
Colorado River Aqueduct
PUMPING STATIONS
Mark Wilmer Pumping Station
Various Stations Along The
CAP Aqueduct
RECHARGE PROJECTS
Various Points Along The CAP
Aqueduct

"first come, first served" system that applies to both the surface water and the groundwater tributaries to a surface stream. The significance of this law is twofold. First, water can be traded as a commodity, independent of the land ownership. Second, these water rights have a fluctuating value.

The critical importance of water in this arid region, and its increasing value, has led to the practice of "water banking," a method of conserving unused river water as a buffer against the shortages caused by the regular droughts. Each year, the Arizona Water Banking Authority (ABWA) pays the delivery and storage costs of bringing Colorado River water into central and southern Arizona through the Central Arizona Project canal. The water is stored underground in existing aquifers (direct recharge) or is used by irrigation districts in lieu of pumping groundwater (indirect or in-lieu recharge). For each acre-foot of water stored, the ABWA accrues credit that can be redeemed in the future when an irrigation district or water corporation in Arizona or a neighboring state needs this backup water supply.

DESERT AGRICULTURE

Fifteen percent of America's crops and 13 percent of its livestock are produced on the 1.5 million hectares of farmland fed by the Colorado River; 80 to 90 percent of the water taken from the river is consumed by agriculture. Yuma, Arizona, and California's Imperial Valley are two of the nation's most important agricultural regions, though their capacity to supply food across the continent, especially during winter months, must be artificially sustained. Farming here is a billion-dollar industry. Imported water and a long growing season allow two crop cycles each year in the Imperial Valley, a major source of winter fruits and vegetables, cotton, grain, and alfalfa (a hugely water-intensive crop) for US and international markets. Considered one of the most productive agricultural regions in the world, the Valley consumes the vast majority of California's Colorado water allocation, as agreed upon in the 1922 compact. Yuma, the "Lettuce Capital of the World," has 40,000 hectares of agriculture supplying almost 90 percent of America's winter vegetables.

In late 1997, the federal government reduced California's Colorado River allocation, and as a result the Imperial Valley Irrigation District's allocation, by 9 percent, to take effect in 2003. The Bureau of Reclamation had ruled that local farmers were not making sufficient use of the water-saving practices commonly employed in other water-starved regions. Beginning in late 2003, Imperial Valley farmers agreed to idle their winter crops, to fulfill a water trade agreement mandating the transfer of thousands of acre-feet of water to San Diego and the Salton Sea.

The local water distribution network—the Imperial Irrigation District (IID)—comprises more than a hundred canal and pipeline branches. The system also includes check dams, spillways, pumping stations, and water reservoirs, all computer-monitored. The water infrastructure that has extended, and in some cases supplanted, the Colorado River is not only constructed, it is highly automated. The Imperial Valley is sustained by the All American Canal, which was completed by the United States Bureau of Reclamation in 1942. Providing drinking water for nine cities and irrigation for over 200,000 hectares of farmland, it is the largest irrigation canal in the world, carrying up to 740.6 m3 (26,155 cubic feet) per second of water.

MANUFACTURED ECOLOGIES—SALTON SEA

Perhaps the most striking aspect of the water infrastructure sustaining the Southwest is the accelerated rate at which landscapes and ecologies are created, erased, and redefined. An extreme example of this environmental impact is the Salton Sea. During a season of heavy rain in 1905, the Colorado River breached its canal, flooding the Imperial Valley and re-filling an ancient inland sea bed. The resulting body of water was 72 km long and 32 km wide, a 932 km2 sea with 177 km of shoreline. The Salton is endorheic, or terminal (without a natural outlet), and has high evaporation rates. It is officially designated as

AGRICULTURAL TYPOLOGIES

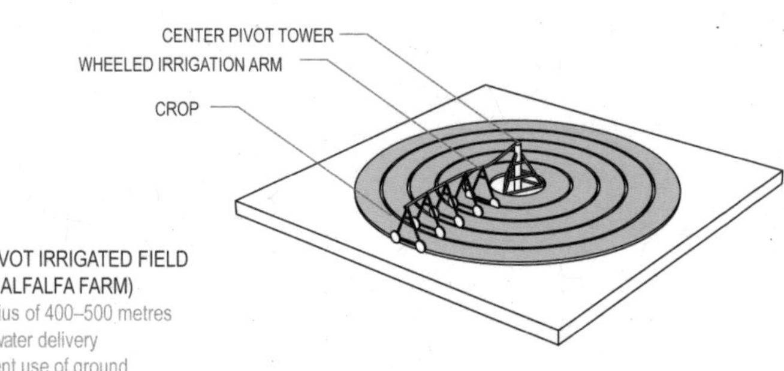

CENTER PIVOT IRRIGATED FIELD
(POSSIBLE ALFALFA FARM)

— circle radius of 400–500 metres
— efficient water delivery
— not efficient use of ground

CROP

RETENTION POND

TRADITIONAL FIELD FARM
WITH RETENTION POND

located north of Salton Sea

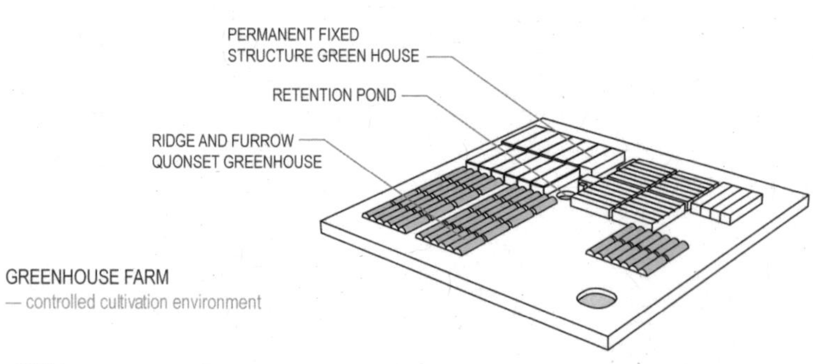

GREENHOUSE FARM

— controlled cultivation environment

HYDROCULTURE TYPOLOGIES

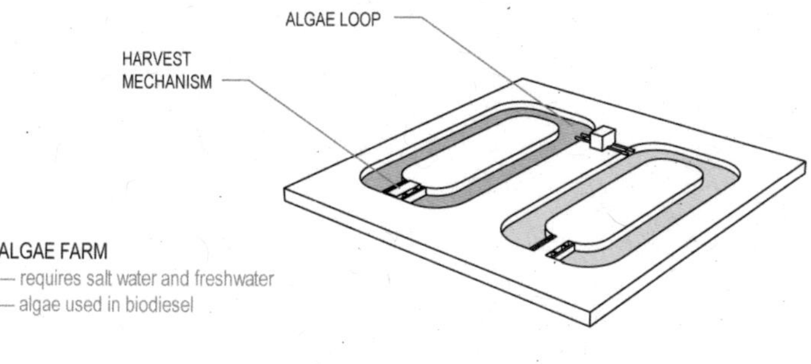

ALGAE FARM
— requires salt water and freshwater
— algae used in biodiesel

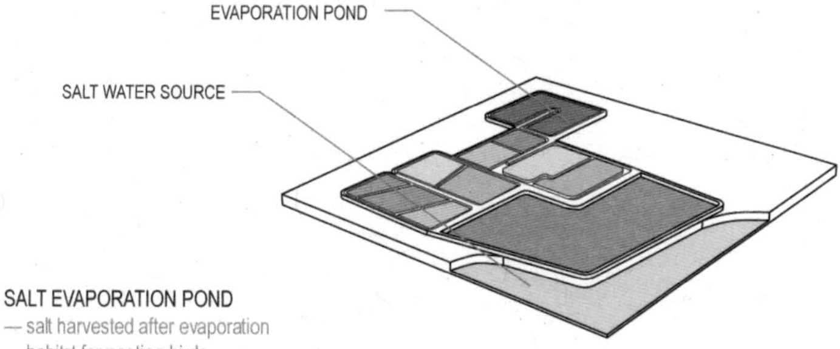

SALT EVAPORATION POND
— salt harvested after evaporation
— habitat for nesting birds

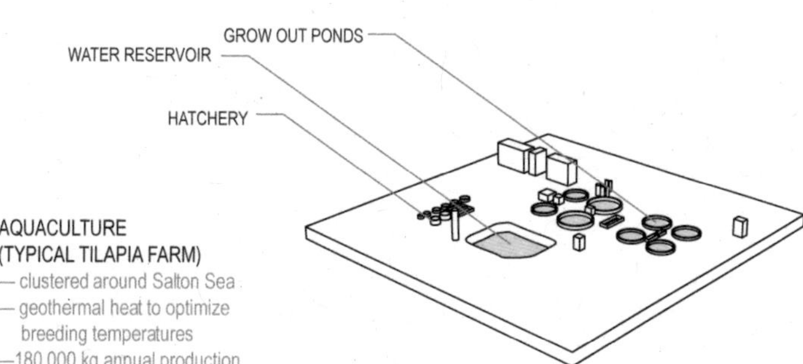

AQUACULTURE (TYPICAL TILAPIA FARM)
— clustered around Salton Sea
— geothermal heat to optimize breeding temperatures
—180,000 kg annual production

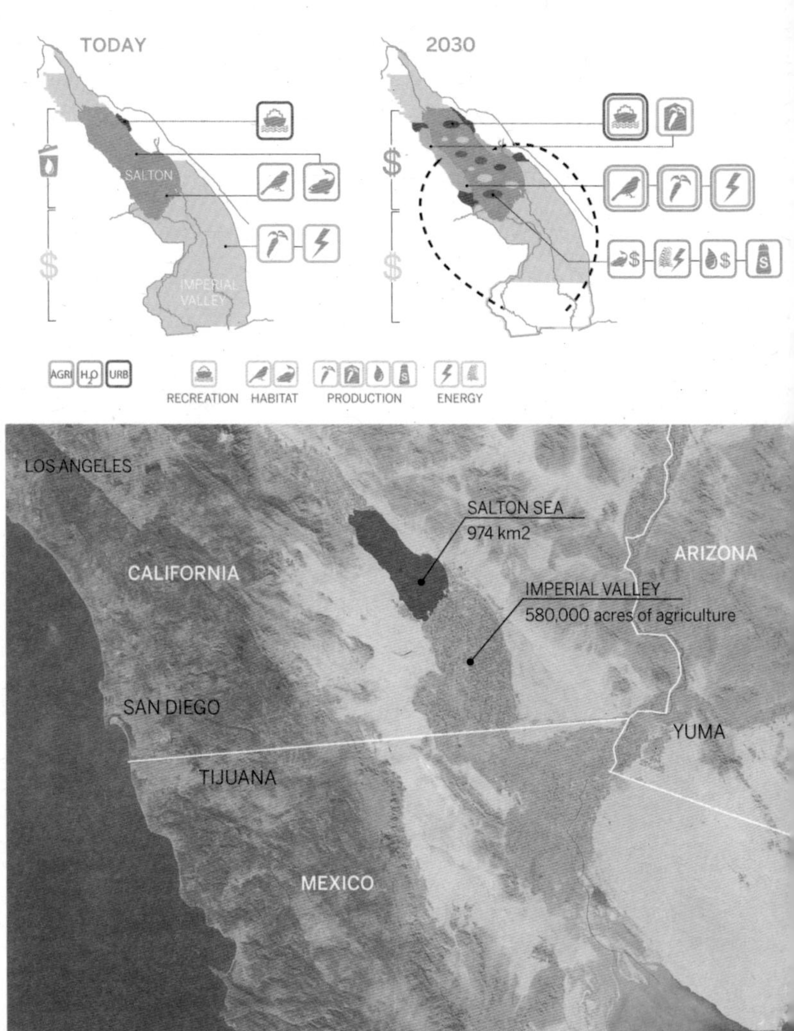
TODAY
2030
SALTON
IMPERIAL VALLEY
$
$
AGRI
URB
RECREATION
HABITAT
PRODUCTION
ENERGY
LOS ANGELES
SALTON SEA
974 km2
CALIFORNIA
ARIZONA
IMPERIAL VALLEY
580,000 acres of agriculture
SAN DIEGO
YUMA
TIJUANA
MEXICO

an agricultural sump for the Imperial Valley, and its water levels are sustained by agricultural run-off. Processes of evaporation and concentration result in a high, and increasing, saline content.

Organized recreation launched the urban development of the Salton Sea area in the 1920s. Boat racing became a major sport there during this period. Because high salinity increases buoyancy, the Salton was known as one of the fastest racing courses in the nation. In the 1950s the California Department of Fish and Game stocked the Salton with a variety of fish species, making it a significant destination for fishing, camping, hiking, bird watching, and boating. Tilapia, for example, a resilient fish that thrives in a high-saline environment, continues to flourish in the Salton. Development peaked in the 1950s and '60s, and the Salton became a California Riviera, with Hollywood film stars enjoying its luxury resorts, country clubs, marinas, and golf courses.

The Salton's fish and bird populations were in dramatic decline from the 1960s to the 1990s, as warm water and agri-chemical pollutants entered the lower food chain species. Many of these species have now bounced back, and by current estimates there are about 80 million tilapia in the Salton Sea. With large numbers of fish come large numbers of migrating birds (nearly 400 species), for whom the sea represents a key stop-over along a Pacific flyway that has been considerably degraded by the massive loss of wetlands throughout California.

One of the major problems with all of the Colorado River's water, exacerbated in the Salton Sea, is salinity. Salts run naturally off soils and rocks, and when river water is used for irrigation, some water evaporates, concentrating salts in the water that returns to the river. Salinity levels are also affected by evaporation from reservoir and aqueduct surfaces, and water use by plants along the river. Currently, the concentration of salt in the Salton Sea is 25 percent more saline than the ocean and increasing by a rate of approximately 1 percent annually, due largely to evaporation. The concentration of salt in the water of the Alamo and New rivers, Colorado tributaries that feed the Salton, is at 44,000 mg/L, or approximately 4.4 percent. By way of comparison, the Pacific Ocean is approximately 3.5 percent salt content, and the Colorado River water, prior to its diversion into the Imperial Valley, is about 0.7 percent salt content.

In 2003, a federally mandated requirement to transfer over 10 percent of the Imperial Valley's water allocation to San Diego threatened the Salton's future. Massive infrastructure projects are proposed to preserve the sea's ecological stability. California has until 2018 to come up with a long-term restoration plan for the Salton; without such a plan the sea will decline rapidly, losing roughly 60 percent of its volume, tripling its salinity, and exposing nearly 300 km2 of lakebed within a dozen years. All of this will lead to massive die-offs of birds and fish as they are displaced from their habitats.

LANDSCAPES ON LIFE SUPPORT

Much as the New Deal infrastructure projects and the federal highway system fundamentally reconfigured the North American landscape in the 1950s and '60s, so national water infrastructure is poised to redefine our notion of natural landscapes in the early twenty-first century. However, the urgent need for these infrastructural landscapes also indicates that there is little room for visions of a nostalgic premodern natural condition in the Southwest. There is no turning back: if these ecosystems and habitats are to be maintained, and agricultural productivity to continue, the landscape must exist on some kind of life support. The question is not if it will require this work or when, but what it will address and how it will be occupied. The Salton Sea and its region of influence are in need of an urgent intervention that manages water in a new way. The iconic dams and canals of the past century are no longer the appropriate design response.

A new approach is needed, but most of the proposals for mitigating the collapse of the Salton Sea rely on old technological fixes: suggestions have ranged from piping water from the Pacific Gulf, with attendant large-scale desalination, to building huge barrier dams, pumping stations, and canals to partition the sea into northern and southern ecosystems. Most projects suggest that maintaining the sea in its current form is impossible, and take the current environmental problems as a liability to be accepted and accommodated. The Salton essentially is viewed as the hydrological refuse point for the farm systems of the Imperial Valley. How can the partitioning, banking, and trading of water in the American Southwest manage such divergent interests? And, instead of being static and engineered, how can water infrastructure be interactive, public, and adaptable?

WATER FARMING: THE SALTON SEA

The Salton Sea's extreme salinity and threatened ecosystems offer an opportunity for economic, social, and ecological innovation. The issue of water demand across the region is central to the sea's future. With urban areas competing for Imperial Valley's allocated use of Colorado River water, the best strategy would be to remediate the water before it enters the Salton.

Our proposal establishes the Salton Sea as a site for water harvesting. In addition, no longer serving as an irrigation sump or supporting a mono-agricultural landscape, the Salton would offer harvests of kelp, algae, fish, and (in a return to its namesake) salt.

In this plan, solar ponds are staged along the Alamo and New rivers, and irrigation water entering the Salton is regulated at ocean-level salinity, or 3.5 percent salt content. Then, rather than a single partitioning of the sea as many have proposed, our plan populates it with floating pools or water-pads of various sizes and salinities. These pools serve as salinity regulation devices as well as farm plots, habitats, and recreational destinations. While other current proposals suggest a more aggressive division of the sea, this scheme envisions maintaining it in its current form, but stabilizing its salinity. The floating pool systems generate micro-ecologies that capitalize on the benefits of higher-salinity, or brine, water, including a rich growth of kelp and

PRODUCTION POOL

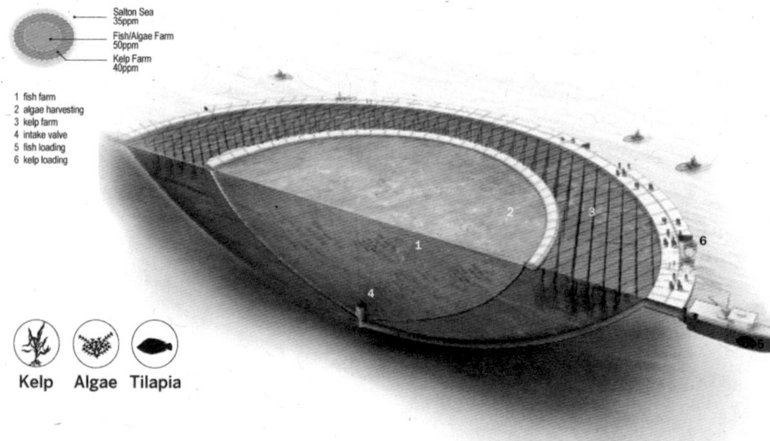

HARVESTING POOL

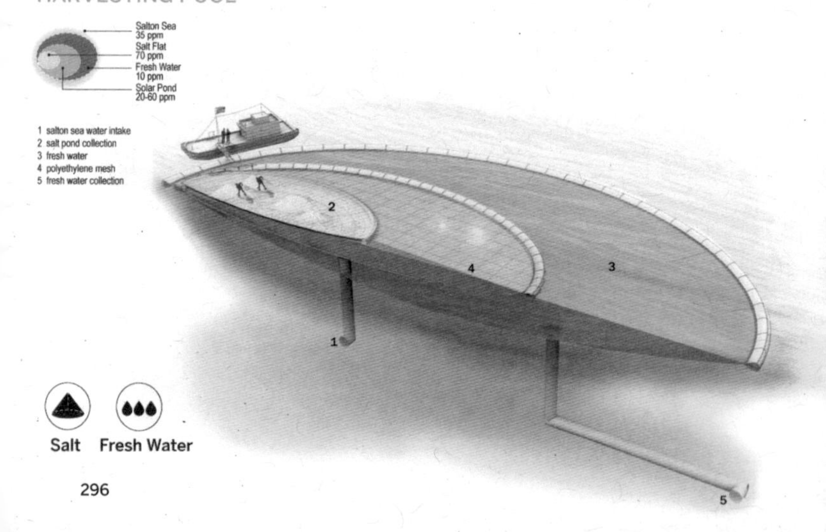

Salton Sea
35 ppm
Salt Flat
70 ppm
Fresh Water
10 ppm
Solar Pond
20-60 ppm

1 salton sea water intake
2 salt pond collection
3 fresh water
4 polyethylene mesh
5 fresh water collection

2
4
3
1
5

Salt
Fresh Water

RECREATION POOL

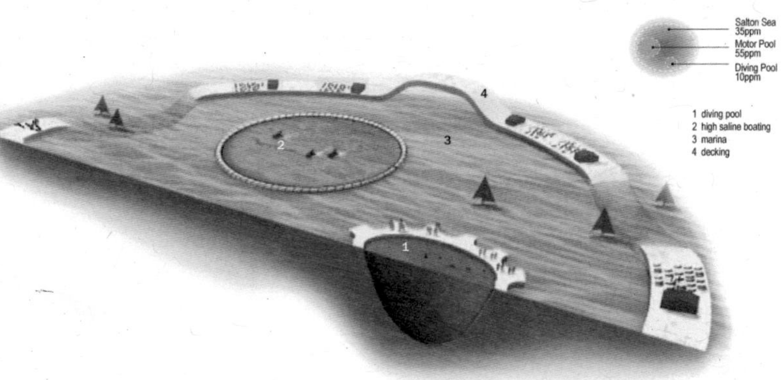

HABITAT POOL

Salton Sea
35ppm

Wetland

Brackish Sea
45ppm

1 fish nursery
2 substrate mesh
3 bird santuary

1

2

3

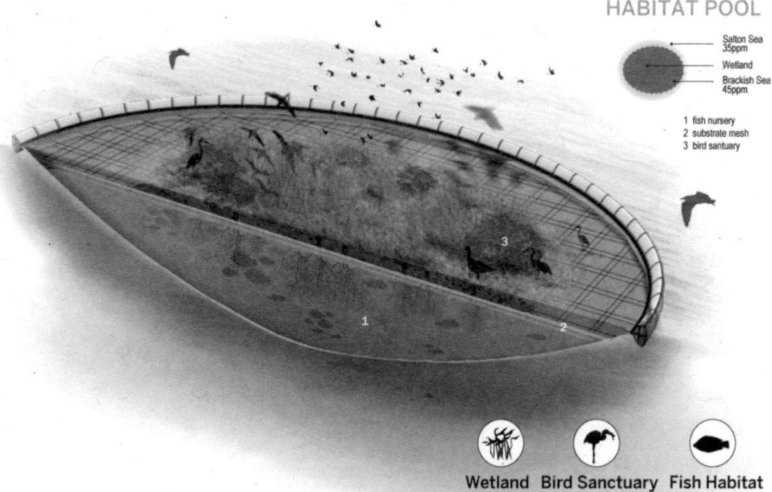

Wetland Bird Sanctuary Fish Habitat

algae, a fertile environment for tilapia farming, and salt crystallization. The next step is freshwater farming. Freshwater harvesting converts ocean saline water into salt crystals and potable water, addressing the water quality issue within the Salton while also generating an economy of water trade for San Diego and Los Angeles.

There are four pool types, varying in scale and complexity, dedicated to production, harvesting, recreation, and habitat. These micro-ecologies are partially moored in place but can also migrate within a territorial range of the Salton. When maintenance or substantive harvesting is necessary, they can be brought close to the shore for collection or upgrade. Along the east and west shoreline of the Salton, the gridded landscape of the Imperial Valley is extended northward to generate a new water-efficient landscape sustained by the Salton and the Coachella Canal. Shifting from traditional water-intensive agriculture, the shoreline farms will integrate sea greenhouses, basins for water treatment and storage, and new wetlands fostering wildlife habitats. A series of bays, jetties, and docks articulate this renewed Salton lakeshore, to support the land-based agriculture and to receive the water-borne islands.

Rather than a New Deal approach of massive engineering or iconic infrastructure, this scheme employs adaptable, responsive, small-scale interventions. Easily replaced or upgraded, these infrastructures double as landscape life support, creating new sites for production or recreation. The ambition is to supplement landscapes at risk rather than overhaul them. The scheme combines existing landscapes with emergent systems to catalyze a network of ecologies and economies in a new public realm.

CREDITS

Design/Research Team: Lola Sheppard and Mason White with Daniel Rabin, Kristin Ross, Valerie Tam, and Joseph Yau.

IMPERIAL VALLEY AGRICULTURE PRODUCTION

PRODUCTION (in US$)

1907 1957 2007 2057

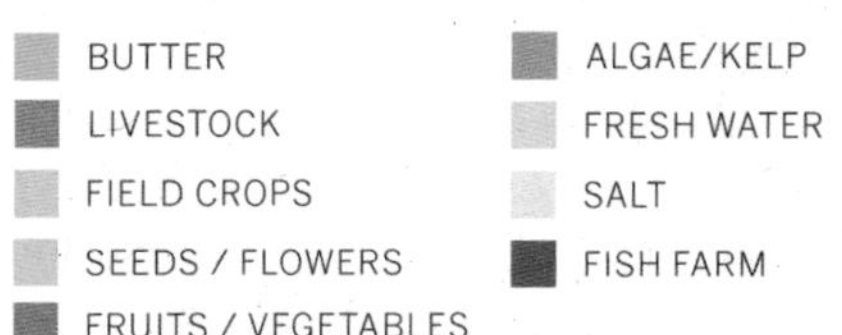

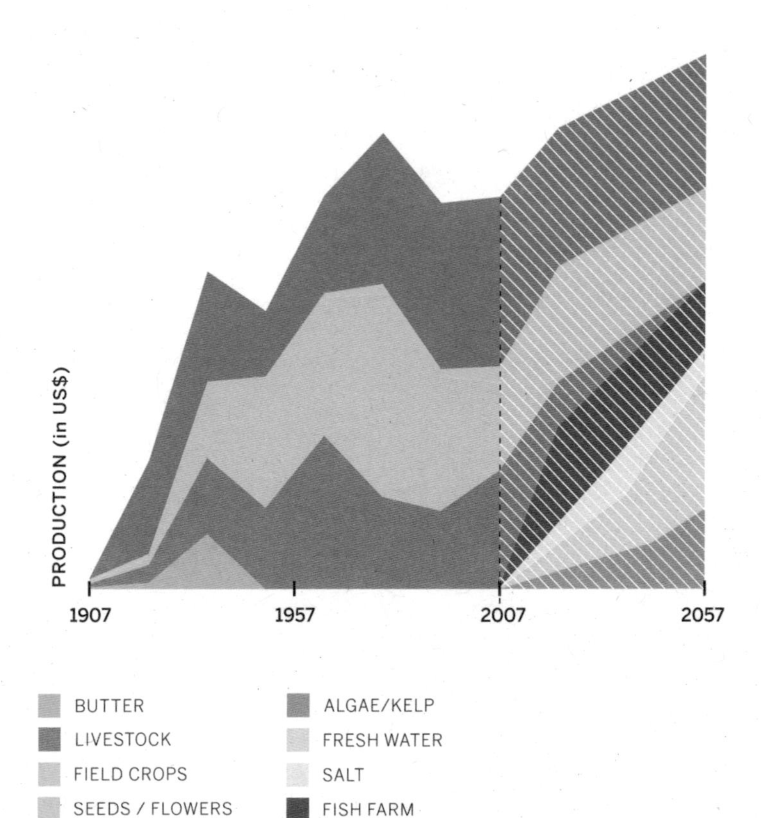

The first 100 years of development in the Imperial Valley has destabilized the landscape. During the next 50 years the valley will capitalize on an aqua-economy and energy harvesting while sustaining this essential regional ecosystem.

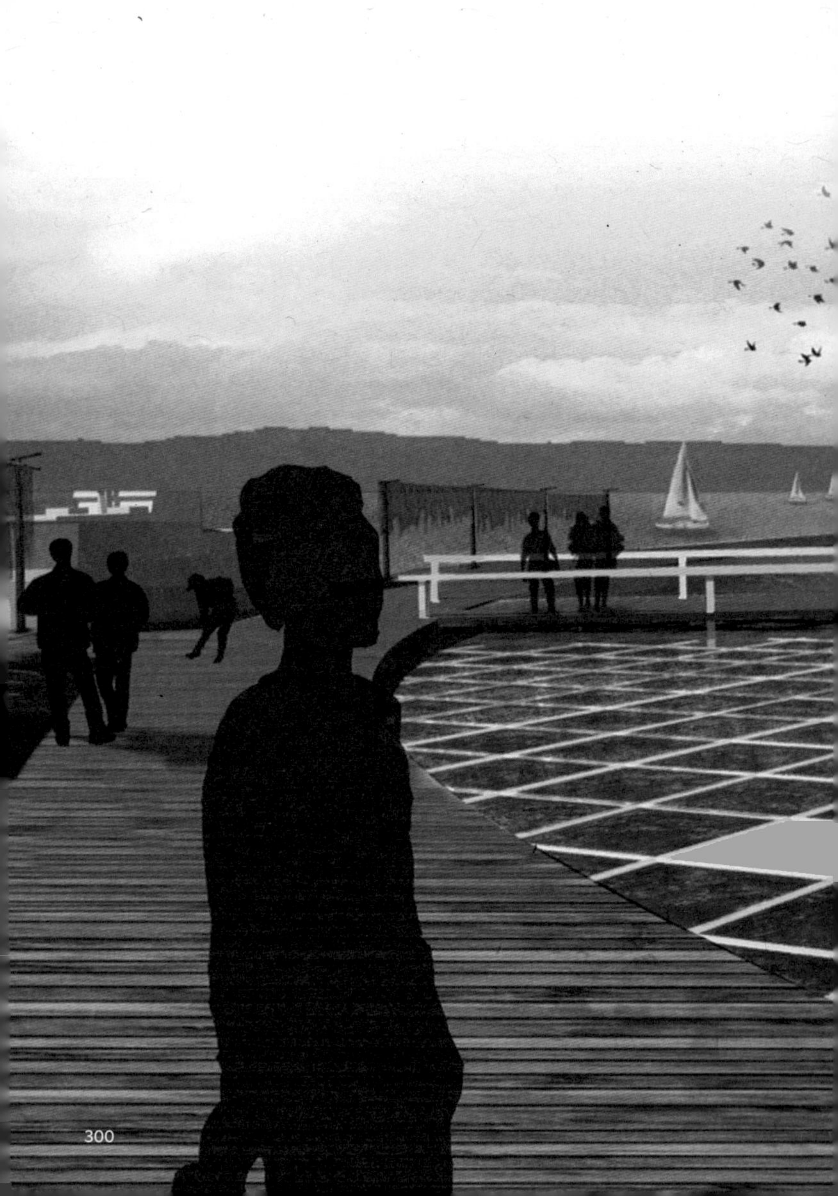

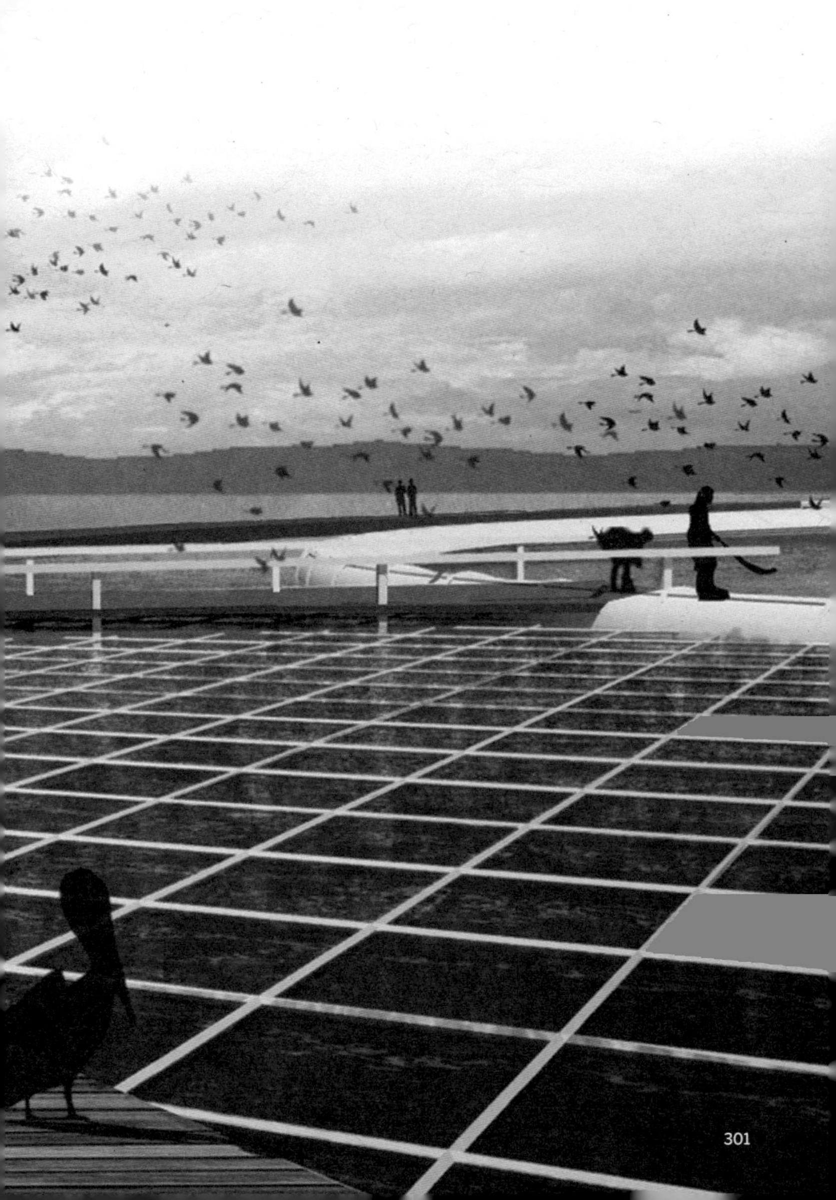

The Price of Clean Water *Bhawani Venkataraman*

The US and other developed nations are fortunate to have easy access to a safe and seemingly unlimited supply of clean water at an exceptionally low cost to the consumer. In New York City, for example, consumers pay just 0.0008 cents for a liter of tap water. Londoners pay more, but still only about 0.15 US cents per liter.

The story is quite different in other parts of the world. Over a billion people worldwide do not have access to clean, safe water, and each year over 5 million people die from water-related diseases. According to the United Nations the minimum daily requirement of clean, safe water for drinking, washing, cleaning, and bathing is 20 liters per person. However, according to a United Nations Development Programme report titled *Beyond scarcity: Power, poverty and the global water crisis*, as many as 1.1 billion people currently have access to only 5 liters a day of clean water. This is equivalent to one-tenth of the daily average amount per person used in a developed nation to flush a toilet. In the US the average per capita daily consumption of clean water is more than 400 liters a day and in Europe it is more than 200 liters a day.

Why do the US and other developed nations have a seemingly endless supply of inexpensive clean water while the rest of the world struggles to provide this necessity of life? A good part of the answer is that while we may perceive water as being cheap, perhaps even free, it is actually far from that. The perception of low cost in the developed world comes about because most of the expense is covered by government subsidies. Though we pay these subsidies through our tax dollars, they are lumped in with other government expenses, and in effect are invisible to us. Governments in developing countries do not have the necessary resources to provide drinking water for all, and so the burden falls on the population, who themselves cannot afford the expense.

To understand the real cost of producing a cup of clean, safe drinking water, we need to look at the costs associated with the processes and systems that supply water to the citizens of a country like the United States. Ironically, our success in producing clean, inexpensive water ultimately leads to both over-consumption and degradation of the quality of our water, thereby raising its cost.

DEFINING AND ENSURING "SAFE" WATER

According to World Health Organization's "Guidelines for Drinking Water Quality," safe drinking water is water that "does not represent any significant risk to health over a lifetime of consumption, including different sensitivities that may occur between life stages." In the US, two congressional acts set in place the regulations that help govern water quality: the 1972 Federal Water Pollution Act, now known as the Clean Water Act, which regulates the quality of water discharged into natural bodies of water; and the 1974 Safe Drinking Water Act, which regulates the quality of the water we consume. A close look at the scientific, technological, engineering, economic, and political factors that come into play as municipalities and businesses work to comply with these regulations reveals that it is neither simple nor cheap to ensure clean, safe tap water.

THE 1972 CLEAN WATER ACT

The Clean Water Act was passed in response to increasing pressure from citizens concerned about environmental degradation. Oil spills and industrial discharges led to unsafe levels of bacterial content, increased deaths of fish, and even, in 1969, a widely reported (but not unique) incident in which a slick of oily contaminants discharged by local industry caused the Cuyahoga River to catch fire. The Clean Water Act regulates the quality of wastewater that can be discharged into natural bodies of water so as to "restore and maintain the chemical, physical, and biological integrity of our nation's waters." In 1974, the Safe Drinking Water Act was passed by Congress "to reclaim and ensure the purity of the water we consume." Both these acts and subsequent amendments establish the regulatory framework to ensure clean water both for human consumption and for the preservation of natural bodies of water.

THE 1974 SAFE DRINKING WATER ACT

The Safe Drinking Water Act regulates levels of contaminants that can be present in drinking water. The EPA website reveals a long list of regulated contaminants categorized as micro-organisms, disinfectants, disinfectant byproducts, organic chemicals, inorganic chemicals, and radionuclides. The EPA defines a "maximum contaminant level" for each contaminant that it regulates as "the maximum level of a contaminant in drinking water at which no known or anticipated adverse effect on the health of persons would occur, and which allows an adequate margin of safety." The maximum contaminant level (MCL) is measured in milligrams (one-thousandth of a gram) of a specific contaminant in a liter of water.

Scientific studies define what constitutes a contaminant and establish safe levels for each substance thus defined. The studies identify sources of contamination, assess the levels at which they pose a risk to human health, and investigate the technical feasibility of detecting, quantifying, and monitoring this risk. Effects not just on the general population but also sub-populations (like the young or older populations) are studied, as are the effects of regulations on the economy. The EPA establishes a priority list of all potential risks and then through a long, painstaking process that includes more scientific studies (and legal challenges), decides whether to regulate this risk and at what level. There are many examples of risks identified by the scientific community that take years to make it to the regulated list or do not get there at all; others are included in the list but at MCLs higher than perhaps some scientific studies recommend. This is not to suggest laxity in the process, but to emphasize the complex non-scientific factors that come into play.

MCLs are regulated at levels that would be equivalent to a tiny fraction of a grain of rice in a liter of water. For example, levels of cadmium must be less than 0.005 milligrams per liter; atrazine, a component of an herbicide, must be less than 0.003 milligrams per liter; and levels of lead in drinking water must be zero milligrams per liter—essentially, there must be no lead in water. These maximum contaminant levels are constantly re-examined as new studies provide more information. The list of contaminants has grown over the years as more threats are identified and more contaminants from human activity are found in water supplies. There are currently over eighty contaminants on the EPA's list, the majority of which are not natural and result from human activity.

THE ROLES OF THE EPA AND STATE ENVIRONMENTAL AGENCIES

Ensuring compliance with these regulations is the job of the Environmental Protection Agency (EPA). Compliance with the Clean Water Act requires each state to assess the quality of its bodies of water by measuring physical, chemical, and biological parameters, such as levels of nitrates, phosphates, pesticides, and bacteria. The Clean Water Act is primarily intended to maintain the quality of natural bodies of water, but when these are also sources of drinking water, the regulations are influential in that area too. As one can imagine, increases in agricultural and industrial activity directly affect discharges into natural bodies of water and hence affect water quality.

THE PATH FROM SOURCE TO TAP

WATER TREATMENT

The process of ensuring that drinking water quality is in compliance with the standards of the Safe Drinking Water Act is complex. Once a natural body of water has been designated as a source for drinking water, measures are put in place to protect it from contaminants. Then the water from this source is processed at a water treatment plant. As described in a National Research Council report, *Drinking Water: Understanding the Science and Policy behind a Critical Resource*, the following steps are typical of what happens at a water treatment plant:

> *Coagulation:* After screening out large objects from the water, coagulant chemicals are added to cause suspended particles to clump together.
>
> *Sedimentation:* Water moves into quiet sedimentation basins where sediments settle out.
>
> *Filtration:* Water is filtered through sand, membranes, or other materials.
>
> *Disinfection:* Chemical additives, ozone, or ultraviolet light are used for disinfection. Other chemicals or processes may also be used to eliminate specific contaminants, to prevent corrosion of the distribution system, or to prevent tooth decay.

Many chemical contaminants are removed at the filtration stage. The complexity of this procedure depends on the level and types of contaminants present in the source. For example, a body of water near an agricultural system may have higher-than-allowed levels of nitrates or pesticides. The local geology also has an effect. For example, arsenic (while also a contaminant from industrial and agricultural activities) exists naturally in some soils, which can result in high levels of dissolved arsenic in water sources. Industrial contaminants can also leach into the water. Each locality has to assess its own situation, taking into consideration all of the factors that could influence the quality of the local water supply, and designing its water treatment plants accordingly.

TESTING AND ANALYSIS

Before the water from a treatment plant is pumped into the water delivery systems—the miles of pipes underground—it undergoes exhaustive testing. Samples are continually collected and analyzed. There is a whole industry devoted to creating and upgrading the technology that is required to monitor, assess, and quantify the quality of water from water treatment plants. The instrumentation used for these measurements must be highly sensitive, capable of detecting contaminants at extremely low levels, and accurate, so that the measurements can be trusted. The EPA has identified protocols for each contaminant that it regulates; laboratories that perform these assessments must be EPA-certified. The process is detailed and time-consuming, and it relies on the technical skills of highly qualified staff. As one can imagine, it is not cheap.

HOME DELIVERY

After the water is purified in the water treatment plant it is pumped into the system that carries it to homes and businesses, an intricate underground network of pipes more than a million miles long in the US. Ensuring that this infrastructure is solid and not compromised is also critical, complex, and costly. The contaminants that have been so carefully removed must not be allowed to re-enter the water on its way from the treatment plant to people's homes. A 2006 National Academy of Science report, *Drinking Water Distribution Systems: Assessing and Reducing Risks*, identifies weaknesses and concerns in the water delivery system:

> Most regulatory mandates regarding drinking water focus on enforcing water quality standards at the treatment

plant and not within the distribution system. Ideally, there should be no change in the quality of treated water from the time it leaves the treatment plant until the time it is consumed. However, in reality substantial changes can occur to finished water as a result of complex physical, chemical, and biological reactions. Indeed, data on waterborne disease outbreaks, both microbial and chemical, suggest that distribution systems remain a source of contamination that has yet to be fully addressed.

MORE MONITORING AND ANALYSIS

To ensure that the water leaves the pipes as clean as it was at the time of its entry, cities and municipalities monitor and assess the water quality at sites distributed around their region. In the late 1990s New York City installed 892 sampling stations to allow a representative sampling and analysis of the quality of water around the city at a total cost of around $11 million. These sampling stations are enclosed taps from which water can be drawn and sent to labs for analysis. Every week, about 1300 samples are collected across the city and analyzed by laboratories using EPA-approved protocols.

TREATING THE WASTE

The water treatment process does not end when the water is delivered to your tap. The "used" water that leaves residences and industries must be treated and monitored to ensure that it meets the standards set by the Clean Water Act before being discharged into the ground or natural bodies of water. This requires specialized treatment plants, infrastructure, engineering, and management. And after the water that enters these plants has been treated, the waste sludge, in another complex process, must be disposed of safely. Finally, the delivery systems and the wastewater treatment plants alike are plagued by problems inherent in the aging infrastructure.

THE COST OF CLEAN WATER

The real costs of the systems that ensure our water quality are difficult to calculate. Water management is a huge, complex process involving many local, state, and federal agencies. As highlighted on an EPA website on water pricing:

> A large portion of our water and wastewater providers are publicly-owned natural monopolies. As such, water and wastewater finances can be intermingled with other municipal departments so that revenues and costs are shared with other municipal functions. Local governments and state laws can also constrain a publicly-owned utility's rate setting. Current water and wastewater prices may not cover the full and complete costs of past and future capital and operating costs.

Setting up these systems requires billions of dollars of investment, and their maintenance costs hundreds of millions more every year. To highlight just one example, New York City in the 1990s needed to build a water filtration plant that would handle 90 percent of its water. The cost estimate at that time was $6 billion for construction and an annual operating cost of $100 million. Ultimately, the city reduced the expense by instead purchasing land and implementing a land management scheme to better protect and preserve the Catskill–Delaware Reservoir. Nonetheless, the cost of this land management plan was still some $1.5 billion. New York City also faces a $3 billion price tag to build a new filtration plant for water from the Croton Reservoir, to comply with EPA regulations, as well as hundreds of millions in annual operating costs.

Obtaining an estimate of the true cost of clean water is further complicated by the differences in the subsidies that various sectors receive for water treatment. Typically about 70 percent of the water consumed in the US is for agricultural purposes, with about 22 percent for industrial use and 8 percent for residential use. In the US, the agriculture sector receives huge subsidies. In some states, farmers pay far less per liter than residential consumers. While it is true that the quality of water used by the agriculture sector is lower than that for drinking water, many of the contaminants on the EPA's regulated contaminant list arise from agriculture. The presence of these contaminants in drinking water supplies adds to the rising costs associated with water treatment and water quality monitoring and assessment.

The rate that residents in the US pay for their water varies from city to city and municipality to municipality. Typical rates are determined by the need to cover costs associated with operating water treatment and wastewater plants and pumping the water to residents. In some cases some capital costs are also included, but not always. In any event, these water rates do not come close to paying for all of the systems we rely on for access to an uninterrupted supply of clean water; the 0.0008 cents per liter of water that New York City residents pay is absurdly low.

THE PROBLEM WITH "CHEAP" WATER

In the late 1700s Adam Smith coined "the diamond-water paradox": water, which is essential to life, is priced far cheaper than diamonds, which are not. However, because the cost of water is very low, we as a society use excessive amounts of it, thereby compromising its quality while at the same time increasing the cost of maintaining a clean supply.

Recent studies suggest that the US may not be able to continue to support delivery of such large quantities of clean water at such a low price at the tap. For example, the EPA estimates that between 2003 and 2023 there will be a need to invest about $270 billion dollars to upgrade and replace aging infrastructure associated with water management. There are increasing threats to our water supplies—new contaminants such as personal care products, MTBE (an additive to gasoline), perchlorates (a component of rocket fuel), and increasing agricultural discharges. Moreover, climate change models predict decreases in our freshwater supplies as well as a higher likelihood of rainstorms in some areas, which will further complicate and threaten water quality and water management systems. All of these challenges mean that we will have to pay more, just to maintain the water quality we now enjoy.

The question, then, is whether the current pricing structure in the US is sustainable. Among the nations of the Organization for Economic Cooperation and Development, the US on average charges the least for water, yet on a per capita basis it consumes twice as much as Europe. Given rising expenses for maintaining the current systems as well as the expected costs of upgraded and new infrastructure, clearly we must question if such low prices for clean water make sense.

If we price water to reflect the true cost of delivering it from source to tap, how will we ensure that everyone has fair access while at the same time respecting it for the precious resource it is? Many models have been proposed for the fair pricing and delivery of water in ways that would encourage conservation, such as block rates, time-of-day pricing, or seasonal rates. Countries in the European Union are increasingly adopting such pricing structures to reduce water consumption, but so far the US as a whole has not. A few cities, particularly those faced with water shortages, have implemented commendable water conservation strategies. Their successes are showing other cities how to adapt to changing realities.

Another important step is reducing water subsidies to the agriculture sector, which consumes by far the largest percentage of water. To ensure that the cost of a reduction in subsidies is not passed on as increased prices to consumers, proven water-conservation techniques, such as drip agriculture and rain-water harvesting, must be advocated.

An unintended yet grave consequence of our water delivery system is the gradual degradation of the fresh water supply. This is inevitable and occurs because no matter how much filtration and water treatment we apply, the effluent water can never be completely free of human-introduced contaminants. True, many contaminants are removed or reduced, but what about those that remain? Since they are not removed, their concentration grows with each cycle through the system, and ultimately they too may be judged harmful and need to be removed, further increasing treatment costs. Moreover, increasing levels of new contaminants are being found in water sources, introduced by human activity. Many of these substances are currently not on the EPA regulated list and the jury is still out as to the level of the threat they pose. But even before any new regulations are put in place, the scientific studies that are needed to assess the risks will entail considerable expense. Clearly, over-consumption further compromises water quality. A pricing structure that limits consumption must be adopted before the costs of maintaining the current levels of water quality become prohibitive.

MORE THAN A BILLION PEOPLE WORLDWIDE HAVE NO CLEAN WATER: WHAT ABOUT THEM?

The substantial cost of providing clean water is often way beyond the means of the governments of developing economies already faced with the monumental and costly problems of providing sanitation, housing, food, and education for their population. The price tags associated with water supply in developed nations are daunting—not just for the construction and operation of water treatment and wastewater plants but also for the associated regulatory, scientific, engineering, policy, management, and technical systems.

The challenge for many governments in developing nations is that they must allocate funding for subsidies and for delivery; the subsidies are necessary to ensure access to safe water, but this expenditure uses up the funds that could ensure clean, safe water for all. Municipalities cannot afford to provide piped water access to all residents, and disproportionately the poor are the ones with the least access to piped water. The poor are invariably faced with two inadequate choices:

either pay the high cost of buying clean water, often at five times the price that a resident with access to piped water pays, or risk using water from local rivers and lakes, which are often contaminated.

Lack of infrastructure is also a huge challenge to the rapidly expanding economies of countries like China and India. Environmental regulations, where they exist, are often not strictly enforced; in transitioning economies this results in increased pollution of water supplies and, if not addressed, quickly will have significant consequences. Environmental regulations must be put in place even as economies expand. The cost of addressing this later, after the damage has been done, will be significantly higher—not just monetarily but also in terms of human health and capital. This is already evidenced by data that reveals increased cancer rates in parts of China where the local population draws water from sources contaminated by industrial waste.

If the possibility of developing economies emulating the water delivery systems of developed nations is unrealistic, at least in the near term, there are nonetheless steps these governments can take, and are taking, to at least improve the situation for the billion or more people who are under their jurisdiction. Improved sanitation and implementing local solutions are two of these steps.

THE BENEFITS OF IMPROVED SANITATION

Inadequate sanitation is the primary cause of the over 5 million water-related deaths that occur yearly. Contamination of ground water supplies is often bacterial due to contact with human and animal feces. Hence, the level of water treatment required and the cost of providing that treatment can be reduced by improving and upgrading sanitation facilities so that the water supplies are not contaminated by waste. These measures provide huge benefits. The associated costs are high, but they are substantially less than the costs of purifying the waste-contaminated water. As stated in a 2008 World Health Organization report titled *Safer Water, Better Health*, "An important share of the total burden of disease worldwide—around 10%—could be prevented by improvements related to drinking-water, sanitation, hygiene and water resource management."

LOCAL SOLUTIONS

It is also important to recognize that for many communities the solutions to these problems do not have to be the same as those used in developed nations. In fact, solutions tailored to the needs and resources of such communities, helping them establish and maintain their own clean water supplies, may actually be more sustainable, and cost significantly less.

One such example is the rainwater harvesting project established by the Center for Science and the Environment (CSE) in New Delhi, India, which helps rural and urban communities take advantage of the torrential monsoon-season downpours. Rainwater is captured in tanks and ponds and also channeled to recharge ground water systems for agriculture and drinking water supplies. CSE works with local communities, running training sessions on water management, including wastewater treatment and reuse.

WaterAID, an international non-profit organization, helps set up water systems for communities in developing countries, also advising and educating them on how improve sanitation and hygiene, focusing on sanitation systems (such as latrines) designed to answer the concerns of the particular community and its traditions.

Researchers have also been developing cheap, simple, and effective water purification systems that can work at the local community level. One example is solar disinfection, or SODIS, which involves exposing water in polyethylene terephthalate (PET) bottles to sunlight for a period of time. Research groups at the Swiss Federal Institute of Aquatic Science and Technology and the Royal College of Surgeons in Ireland have demonstrated this method of disinfection to be effective in killing many of the bacteria that cause water-borne diseases. Field studies have shown a 30 to 80 percent reduction in diarrhea cases for people consuming SODIS-treated water. While this method does not remove chemical contaminants, or all biological contaminants, it can have dramatic positive effects in places where bacterial contamination in water supplies is the primary concern. Moreover, the simplicity of the system allows it to be easily set up in any rural community.

THE BENEFITS OF PROVIDING CLEAN WATER

When clean water is unavailable, societies suffer. Studies show that access to clean water reduces child mortality and the spread of infectious diseases, improves maternal health and greatly enhances quality of life for the poor. These benefits also strengthen human resources by increasing economic equity, creating educational opportunities for girls, empowering women, and, through improved agricultural production, reducing hunger.

Under Goal 7 of the UN Millennium Development Goals (MDG), "Ensure Environmental Sustainability," is the following target: "Reduce by half the proportion of people without sustainable access to safe drinking water and basic sanitation." A study commissioned by the World Health Organization suggests that investments in meeting this MDG goal will result in economic benefits ranging from $5 to $60 per US dollar invested depending on the geographic region. The price for achieving this goal will be significant. A 2008 study by Guy Hutton and Jamie Bartram in *The Bulletin of the World Health Organization* estimated that to meet the MDG target by 2015, US$ 184 billion will be needed for new infrastructure, with an additional $538 billion required to maintain existing services between now and 2025. However, these investments will more than pay for themselves in terms of human development. In fact, not investing in meeting this goal would add enormous costs to the current clean water delivery system in these areas. As a point of reference, the global annual expenditure on bottled water is estimated to be $100 billion.

THE ROLE OF EDUCATION

Investments in education at all levels will also help our understanding of the true cost of clean water. Education can promote understanding that clean water should not be taken for granted or frivolously consumed, and that our sources of freshwater must be protected. Just because water appears to be cheap does not mean it should not be valued as if it were gold—or diamonds.

In the US, the EPA requires cities and municipalities to publish water quality reports to inform local citizens about their drinking water. But in reality, how many of us read these reports? Perhaps if we did we would be pleasantly surprised at the high quality of our tap water, and be less inclined to see bottled water as a superior alternative. (In fact, in spite of its high cost relative to tap water, bottled water is not as stringently regulated.)

A convincing public-service ad campaign to educate the public could encourage conservation. If the true cost of the water we consume were understood by all sectors, perhaps we would be more ready to support the rising, but essential, costs of upgrading our aging infrastructure—something that must be done before it is too late. Other successful conservation strategies, such as grey water and rainwater harvesting, need to be promoted; such measures could be required in any new building designs and for agricultural and industrial use.

Water, once consumed, returns to the hydrological cycle; that will not change. But after human use it is not of the same quality as before. It is therefore imperative that we begin to implement strategies that move us toward conservation. After all, as the saying has it (in many languages), water is life.

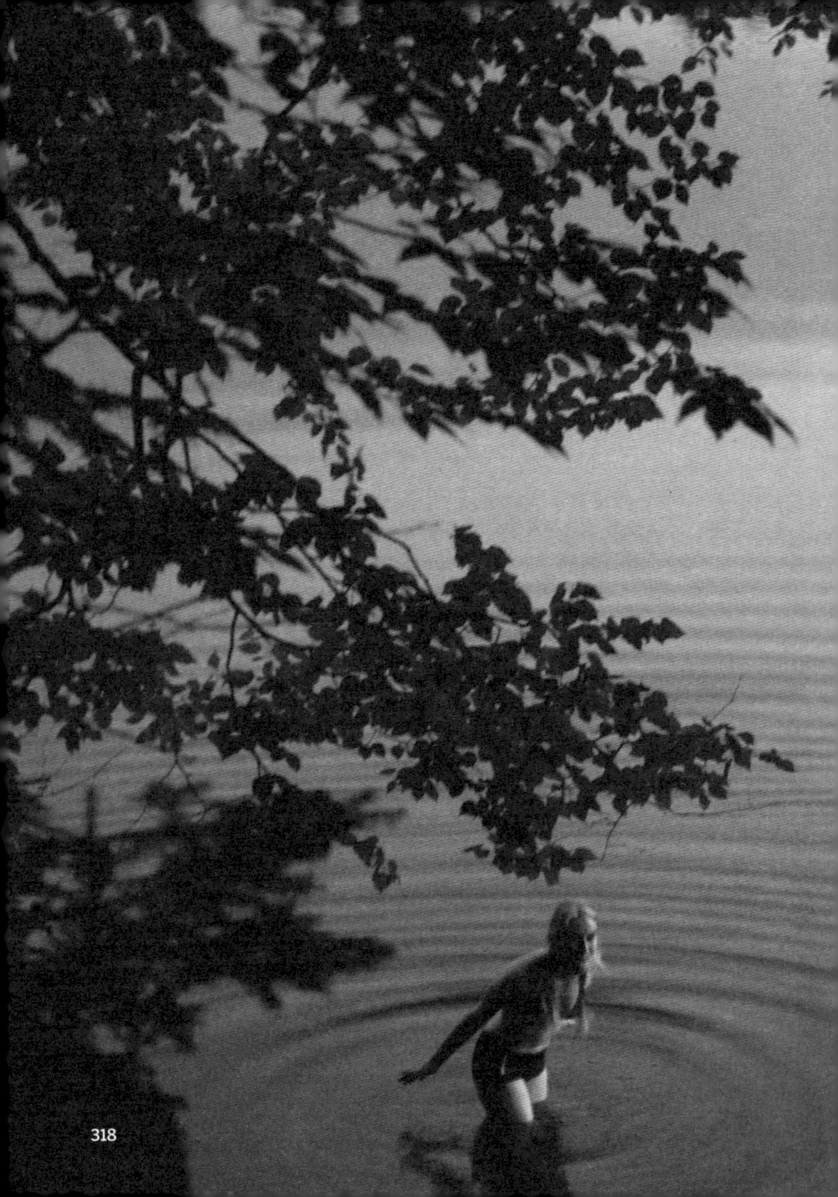

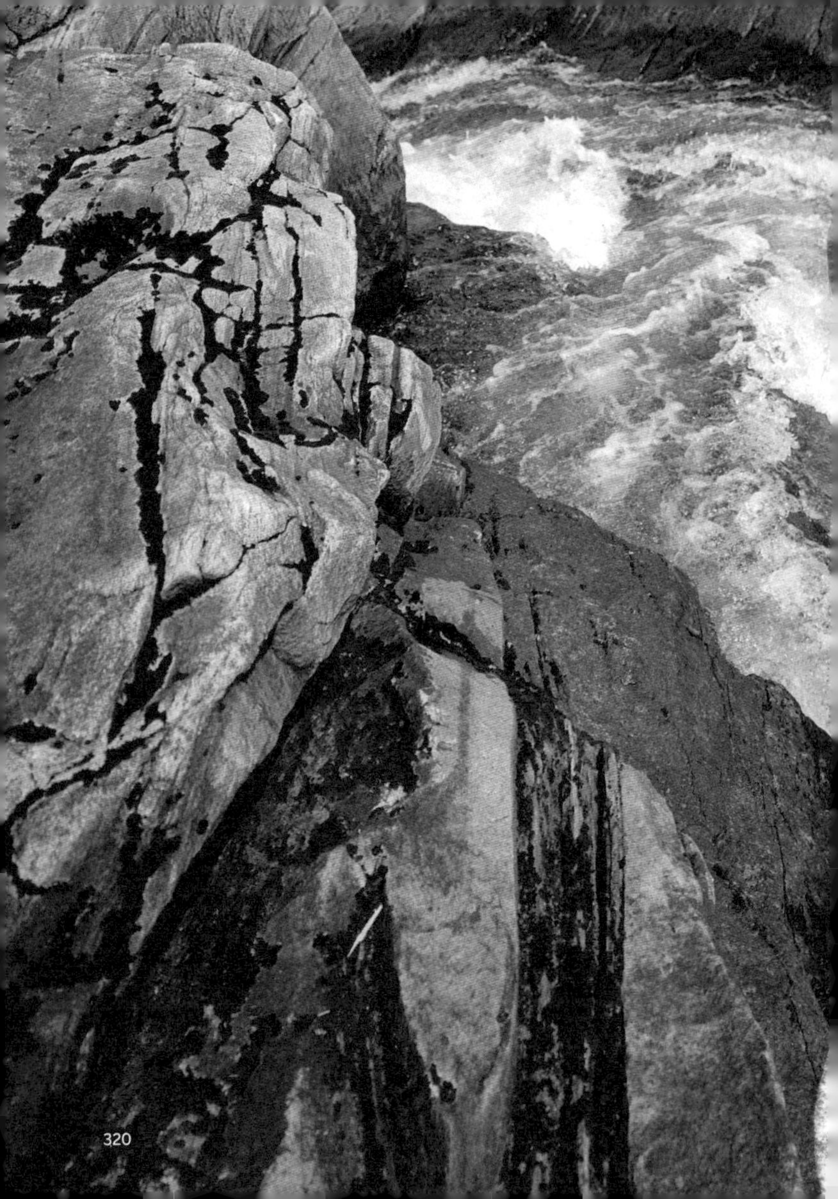

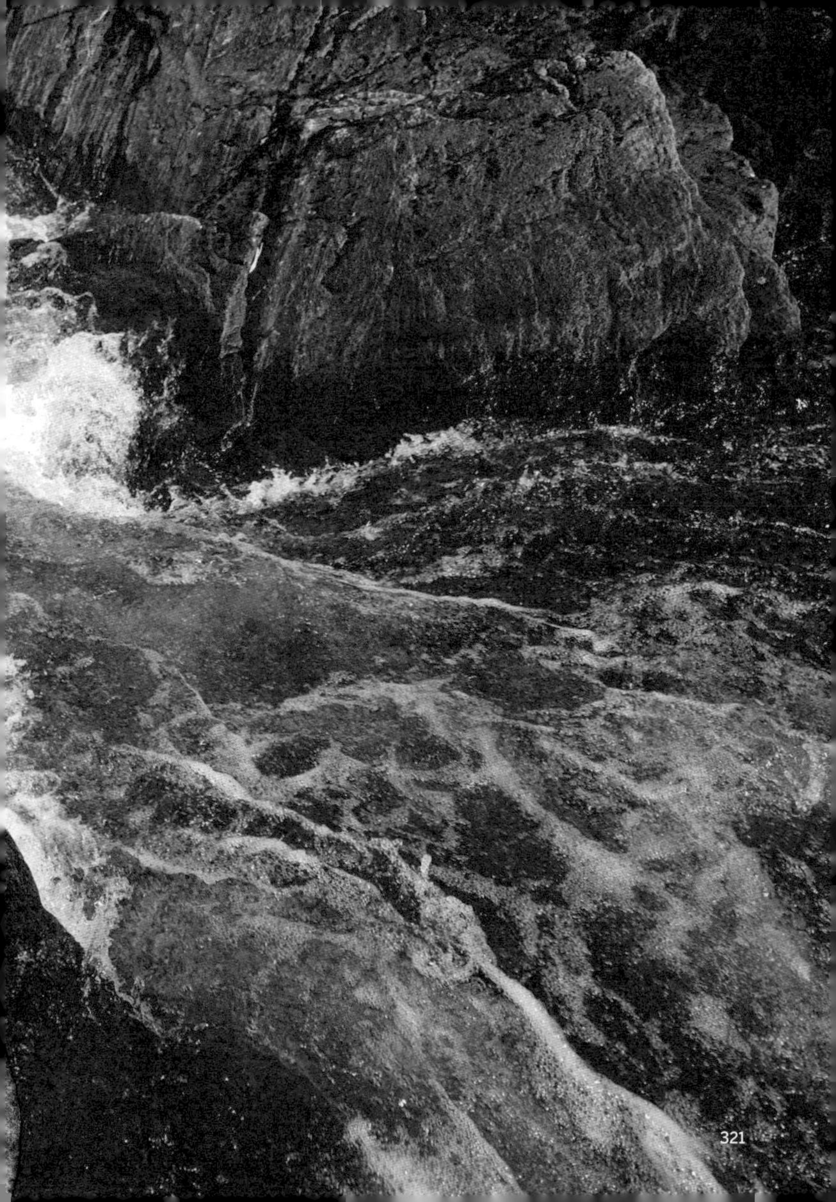

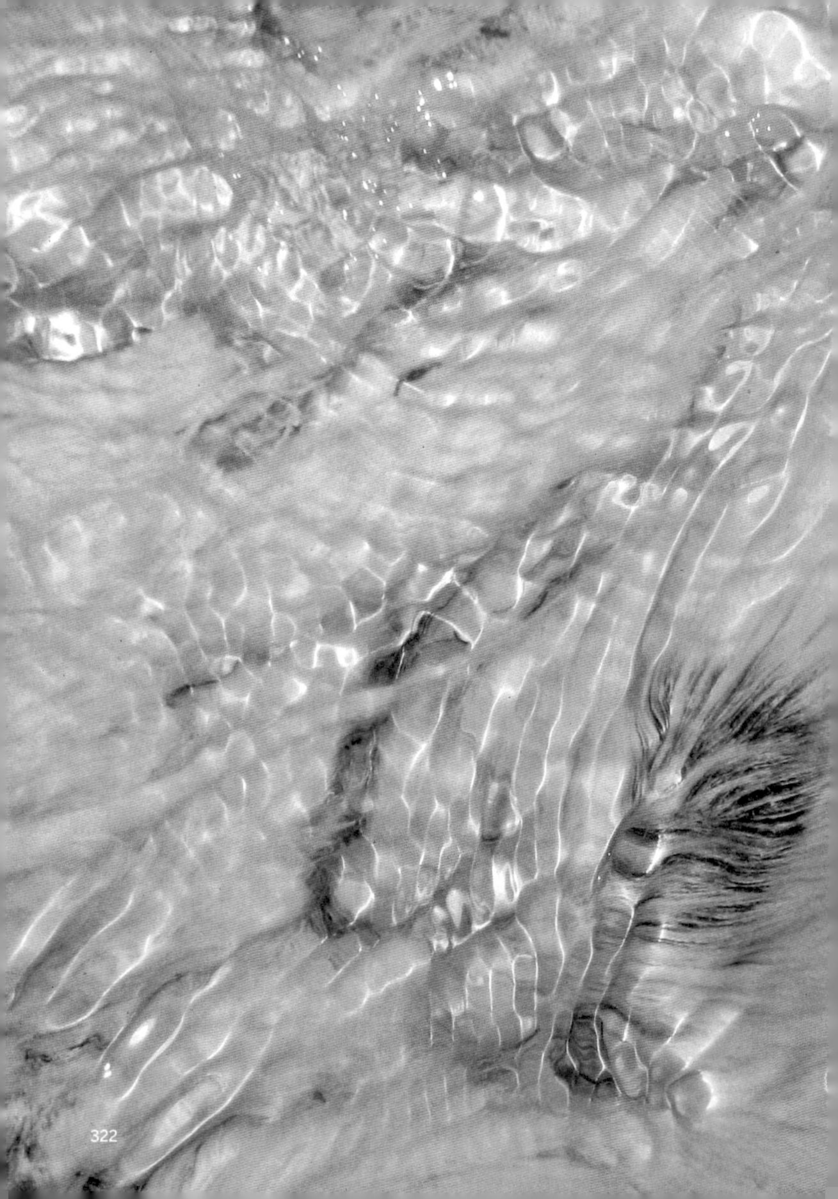

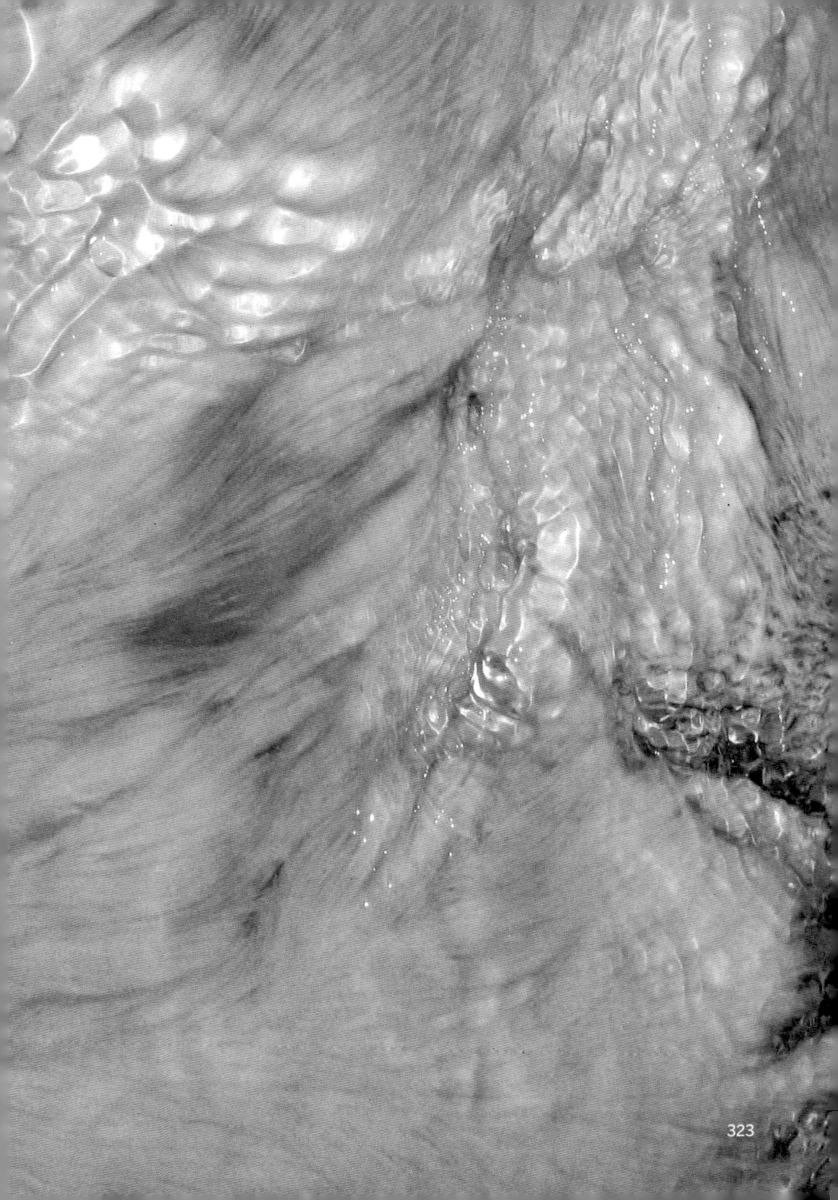

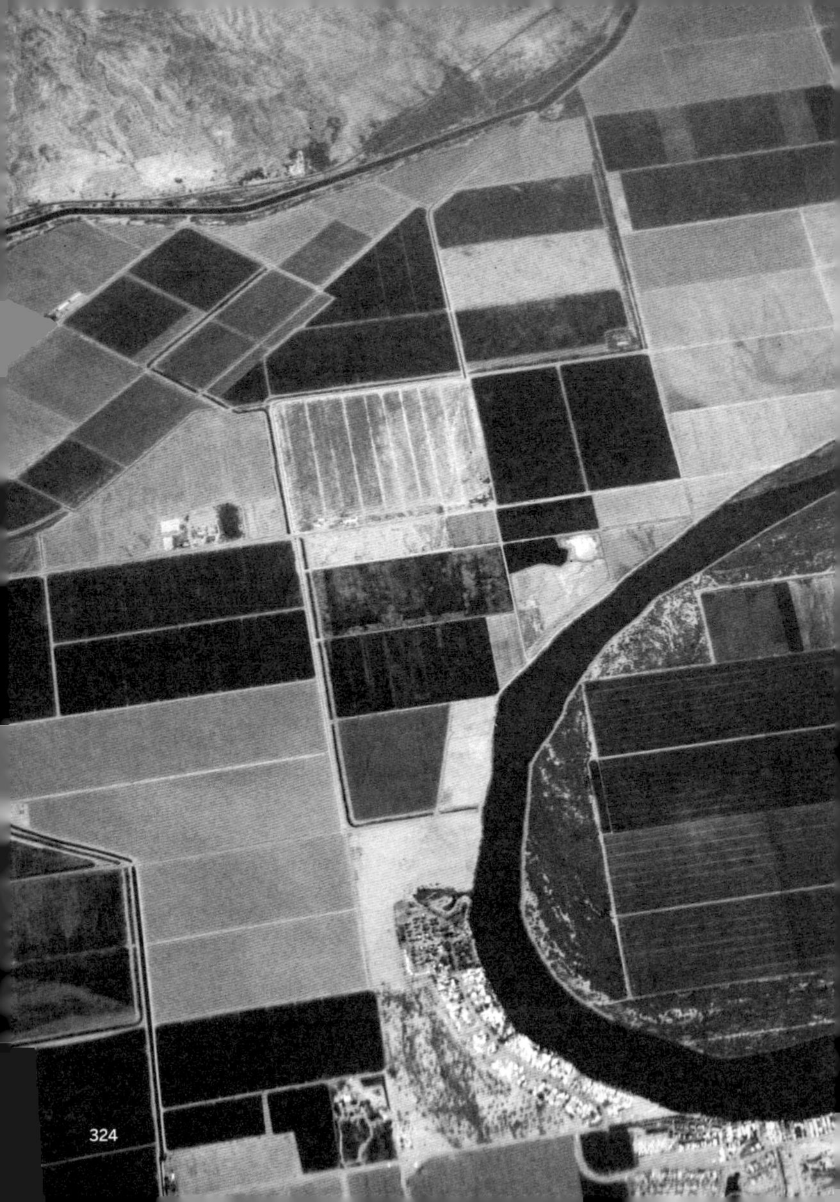

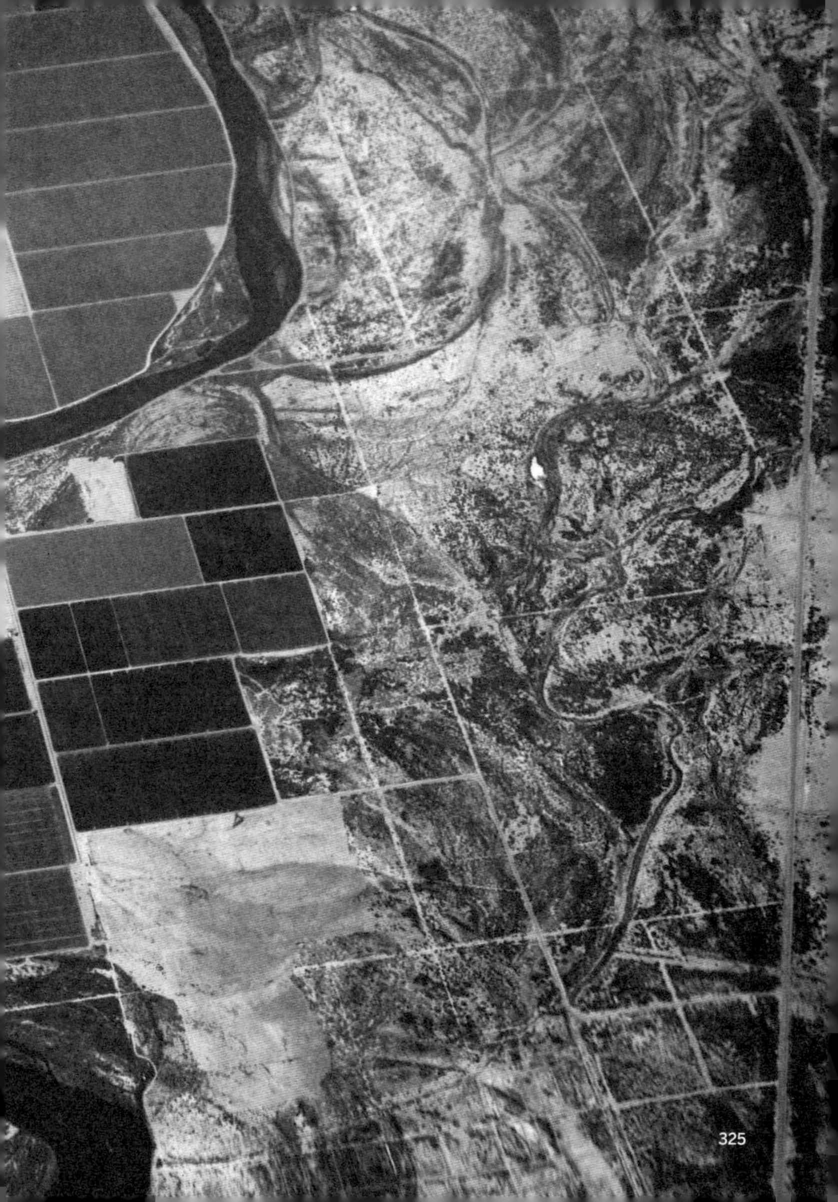

WARREN'S WATERLESS PRINTING

Canada's leading environmental printer using high-quality waterless print technology. This project was printed by Warren's Waterless Printing using the waterless printing process. Warren's has reached beyond the standard environmental practices to ensure it delivers the most environmentally friendly and high quality printing outcomes available. The waterless printing process eliminates the use of fresh water and greatly reduces the use of harmful chemical compounds. This ensures that harmful chemicals are not emitted into the environment, and that water is not wasted.

ENVIRONMENTALLY FRIENDLY PAPER (RECYCLED PAPER)

The cover of this project was printed on FLO paper which contains 10% post consumer fibre. It is elemental chlorine-free and is FSC certified. The text was printed on Rolland Enviro 100 which contains 100% post consumer fibre and is FSC certified.

FOREST STEWARDSHIP COUNCIL (FSC) CERTIFICATION

This project was printed on paper and by a printer that is Forest Stewardship Council (FSC) certified. FSC promotes environmentally appropriate, socially beneficial and economically viable management of the world's forests.

BULLFROG POWER — CLEAN, RELIABLE ELECTRICITY

This project was produced utilizing Bullfrog Power to power the printing presses. Bullfrog Power generates power exclusively from wind and low-impact water power generators, and meet or exceed the federal government's EcoLogo standards for renewable electricity.

TERRACHOICE–ECOLOGO CERTIFICATION
ENVIRONMENTAL CERTIFICATION BY THE GOVERNMENT OF CANADA

This project was printed by an EcoLogo certified printer. This program recognizes manufacturers and suppliers of environmentally preferable products and services.

ISO 14001

This project was produced by an ISO 14001 certified printer. Fully committed to continuous improvements in waste reduction and pollution prevention, Warren's has a comprehensive environmental management system that integrates sound business practices with environmental, health, safety and quality control practices.

ENVIRONMENTAL PRINTING SUMMARY

The following summarizes the state-of-the-art environmental practices that have been incorporated into this project.

ECO-CALCULATOR

Using 4,584 lb. of Rolland Enviro100 Print instead of virgin fibres paper reduces your ecological footprint of:

Tree(s): 39
Solid Waste: 1,123 KG
Water: 106,239 L
Suspended particles in the water: 7.1 KG
Air Emissions: 2,466 KG
Natural Gas: 160 M3

It is the equivalent of:
Tree(s): 0.8 American football field(s)
Water: a shower of 4.9 day(s)
Air Emissions: emissions of 0.5 car(s) per year

Paper partially donated by Cascades Fine Papers Group

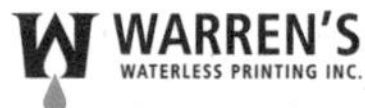

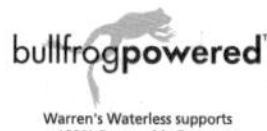

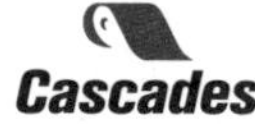

Recycled
Supporting responsible use of forest resources
www.fsc.org Cert no. SGS-COC-005833
© 1996 Forest Stewardship Council

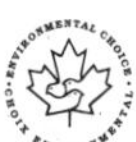

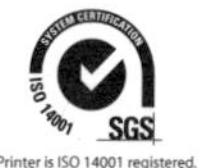

Printer is ISO 14001 registered.

EDITOR
John Knechtel

MANAGING EDITOR
Jennifer Harris

ART DIRECTOR
Studio:Blackwell
Kelsey Blackwell with Judith McKay

COMMISSIONING EDITORS
Jocelyne Alloucherie
Stephen Andrews
Rita Bakacs
Pierre Bélanger
Adrian Blackwell
Jessica Blaustein
Daniel Borins
Heather Cameron
Mei Chin
Mark Clamen
Karen Connelly
Roger Conover
Rebecca Duclos
Atom Egoyan
Kevin Finlayson
Camilla Gibb
Dorothy Graham
Janna Graham
Angela Grauerholz
Melissa Grey
John Greyson
Megan Griffith-Greene
Chris Hardwicke
Terrance Houle
Mark Kingwell
Robert Kirkbride
Mark Lanctôt
Gilbert Li
Pamila Matharu
Doina Popescu
Lisa Rochon
Joseph Rosen
David Ross
Ann Elisabeth Samson
Jim Shedden
Ann Shin
Timothy Stock
Kevin Temple
Sonali Thakkar
Greg Van Alstyne
Bhawani Venkataraman
Mason White
Ger J. Z. Zielinski

COPY EDITOR
Doris Cowan

PREPRESS
Clarity

PRINTER
Warren's Waterless, Toronto

BOOK BINDING
York Bookbindery, Toronto

LEGAL REPRESENTATION
Caspar Sinnige

CONTACT
alphabet-city.org
director@alphabet-city.org

UPCOMING ISSUES

FALL 2010
no. 15: AIR

FALL 2011
no. 16: SPORTS

FALL 2012
no. 17: SCHOOL

FALL 2013
no. 18: CHINA

WATER WAS PUBLISHED WITH THE SUPPORT OF

Canada Council for the Arts
Conseil des Arts du Canada

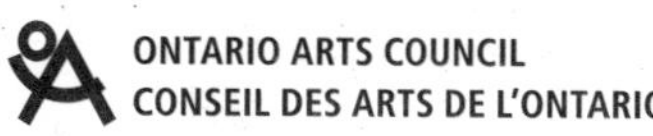

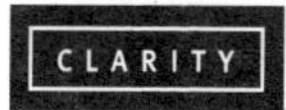

DRAKE HOTEL

EAMON MAC MAHON PHOTOGRAPHIC CREDITS

Inside front endpapers: *Swimming Pool, Mississauga*
Following Spread: *Llewellyn Glacier, British Columbia*
pp. 2–3: *Shore Break, Haida Gwaii*
pp. 4–5: *Gated Community, San Francisco Bay Area*
pp. 6–7: *Steam Tornado, Hawaii*
pp. 8–9: *Megin River, British Columbia*
pp. 10–11: *Puddle, Haida Gwaii*
pp. 12–13: *Schoolyard, Vancouver*
pp. 18–19: *Inside Passage, British Columbia*
pp. 160–170: *Lake Ice, Yukon Territories*
pp. 318–319: *Mont Tremblant, Quebec*
pp. 320–321: *Moon River, Ontario*
pp. 322–323: *Baranof Hot Springs, Alaska*
pp. 324–325: *View From Jet, Southwestern United States*
Inside back endpapers: *Swimming Pool, Mississauga*

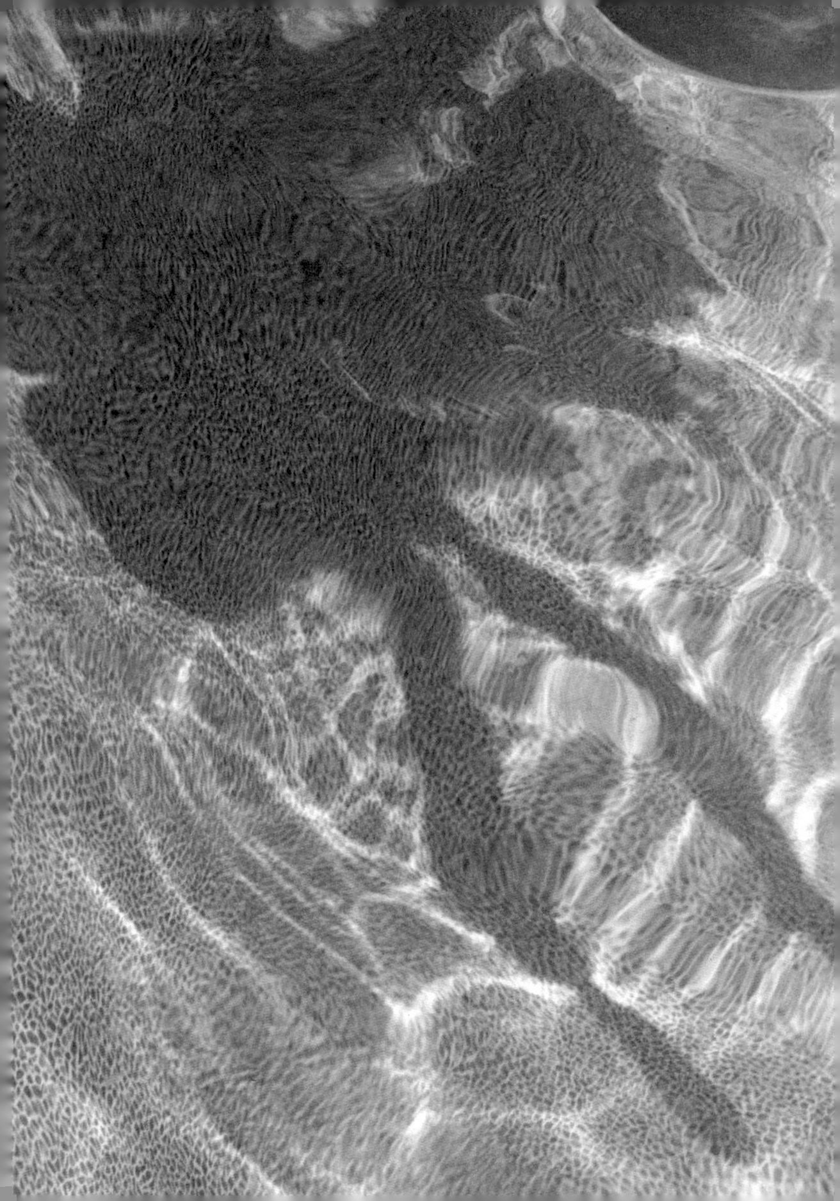